Praise for
BRIDGING SCIENCE AN

D0150099

"*Bridging Science and Spirit* correctly asks, 'How does matter originate from consciousness?' This is the fundamental question of a growing body of literature regarding the new paradigm of an idealist, consciousness-based science. Norman Friedman has made an important and thoughtful contribution to this new science."

Amit Goswami, Ph.D. — Professor of Physics, University of Oregon. Author of *The Self-Aware Universe*

"Some future historian will, I feel confident, identify the 'Great Debate' of the twentieth century around the question: 'What is science going to do about consciousness?' I would surmise that Norman Friedman's book *Bridging Science and Spirit* will turn out to be a benchmark in that inquiry... No one can read it without gaining some clarity on their own nature."

Willis Harman, Ph.D. — President, Institute of Noetic Sciences. Author of *Global Mind Change*

"Few have understood both science and metaphysics well enough to remove our blinders to their underlying similarities—and universal truths. Kudos to Norman Friedman for providing to be a master bridge builder."

Lynda Dahl — President, Seth Network International. Author of *Beyond the Winning Streak; Ten Thousand Whispers; The Wizards of Consciousness*

"...a brilliant synthesis of ageless wisdom and modern science. Friedman's writing provides pictures for our minds so we can 'see' reality...."

Jacquelyn Small, MSSW — President, Eupsychia, Inc. Author of *Awakening in Time; Transformers, the Artists of Self-Creation*

"Norman Friedman's special gift is his ability to bring out the imaginative significance of the new paradigm. Recommended for all who enjoy having their minds expanded."

Michael Grosso — Author of *Frontiers of the Soul; Soulmaker*

"I responded to this book with both admiration and appreciation for the task Norman Friedman set out to do and accomplished so successfully...; He has paved the way toward a deeper understanding of the role of consciousness in the creation and structure of matter."

Montague Ullman, Ph.D. — Clinical Professor of Psychiatry, Emeritus, Albert Einstein College of Medicine, N.Y.

BRIDGING SCIENCE AND SPIRIT

Common Elements
In David Bohm's Physics,
The Perennial Philosophy and Seth

— Norman Friedman —

Living Lake Books

Bridging Science and Spirit

First printing, October 1993
Second printing, June 1994
Third printing, April 1997

Library of Congress Cataloging-in-Publication Data
Friedman, Norman, 1926-
    Bridging science and spirit : common elements in David
Bohm's physics, the perennial philosophy and Seth / Norman
Friedman.
        p. cm.
    Originally published: St. Louis, MO : Living Lake Books,
    c1994.
    Includes bibliographical references and index.
    ISBN 1-889964-07-7
    1. Reality 2. Physics—Philosophy. 3. Metaphysics.
    4. Mysticism. 5. Bohm, David. 6. Wilber, Ken. I. Title.
[QC6.4.R42F75   1997]
113—dc21   97-5767
                                                           CIP

The Woodbridge Group
PO Box 849 ◆ Eugene, OR ◆ 97440 ◆ USA
(541) 683-6731 ◆ fax (541) 683-1084

Dedicated to the memory of David Bohm (1917-1992), whose intellectual honesty in facing the difficult philosophical issues of quantum theory has been deeply inspiring.

Acknowledgments

Leah Friedman, my wife and best friend, spent countless hours reading and correcting the initial manuscript and gave me much needed encouragement throughout the project. Carolyn Fortel, my secretary, patiently typed, retyped, and checked my work, improving it in numerous ways. Sara Jenkins, my editor, managed to both perform her work with skill and enthusiasm and become a good friend in the process. My publicist, Marisa Lovesong, provided valuable assistance and tireless dedication to the success of this project. My children and grandchildren, just by being there and being who they are, contributed greatly. I am grateful to them all.

I also thank Fred Alan Wolf for writing the foreword and his continued encouragement.

Contents

Foreword

When I received a doctorate in theoretical physics, I joined a club of like-minded seekers who dreamed of capturing the Holy Grail of the ultimate knowledge of Life, the Universe, and Everything. We physicists were thoroughly trained in the workings of the universe, from the soupy fluids of thermodynamics in classical physics to subatomic nuts called particles in the farthest corners of space and the tiniest fleeting moments of time.

Yet as we sought truth at every point and at every instant, we found something lacking. Perhaps I should speak only for myself in describing the sudden feeling of emptiness when I first saw that Einsteinian relativity theory was completely incomprehensible as an objective experience. Who would ever go speeding off at the velocity of light? The mere idea was mysterious, to say the least, that as one approached light speed, time would slow down compared to an earthbound observer, and yet be experienced not as a slow-down but simply as taking less time to get wherever one was speeding off to.

The mathematical theory of relativity is actually amazingly elementary, and after wrestling with it for awhile, I eventually grasped the idea. But understanding it in terms that my classical education would allow? That was a "horse-question" of a different color, involving concepts I had never encountered or imagined. Then, with quantum mechanics, suddenly the world was no longer made of tiny particles, fleetingly existing or not, but of mysterious flowing probabilities, which enabled physicists to predict with certainty only the probability of nearly everything, but not the actual occurrence of anything.

Yet the mathematical structure of this "new physics" was compelling. And after much soul-searching and many years of writing, teaching, and problem-solving, I can say that I have come to accept the mystery of the new-physics picture of the universe and the remarkable power it seems to have, based on something quite intangible and seemingly outside our everyday experience: the invisible flow of possibilities that exist in the abstract world of mathematics and intersect with our real world in terms of the prediction of probabilities of real events. It is understandable that I — and perhaps other physicists who write about the new physics — assumed that anyone outside the field would find it entirely baffling.

I held that assumption until I read Norman Friedman's book, the one you are holding in your hands. This book impressed me, to say the least. Here is someone outside the field, looking in, and spotting both the logic and the magic of the discoveries of the last nearly one hundred years of soul-searching on the part of certain physicists, and bringing it together in a clearly written summation.

This soul-searching has led many of us to a new realization, that the gap of understanding separating the two seemingly irreconcilable viewpoints of spirituality and science can be bridged. This idea is admirably presented by Mr. Friedman. Indeed, it is more than a bridge that this author has built; he has shown us that these two approaches aim toward the very same truths. It is as if one were to cross a bridge from one country to another and find that one had arrived right back where one had started.

Reading this book, I was struck with a feeling of satisfaction. I had felt, after writing about such concerns as addressed by Mr. Friedman, that hardly anyone really understands just what the problem is: namely, the outrageous logic that shows that the physical cannot exist independently from the mental, that ontology and epistemology are the same. Certainly the connections between the deep philosophical issues raised by the fact of human consciousness and the seemingly very different viewpoint taken by the new physics seemed much too difficult to be grasped by someone outside the field. I was wrong. Mr. Friedman really understands the problem and has put together a remarkable book that explores the "bridge" with skill and insight.

Bridging Science and Spirit accomplishes a formidable task. Every important facet of the problem has been addressed both with intelligence and with heart. The key areas of overlap that form the basis for Mr. Friedman's insights are in the work of physicist David Bohm, the mystical perspective as elucidated by the writings of Ken Wilber, and the visionary teachings of Seth, the discarnate entity channeled by Jane Roberts.

Remarkably, I first became acquainted with Seth's teachings shortly after a period of study and research at the University of London's Birkbeck College where, as fortune would have it, I occupied an office next to Professor Bohm. Bohm would often tell me about his latest insights into the implicate order, an invisible, pervasive "isness" from which all matter, energy, and meaning ultimately emerge. At the end of these discourses I felt as bewildered as the woman who met Einstein on a ship sailing from Europe to the U.S. and remarked after Einstein had explained his theory of relativity to her, "I am convinced that he understands it very well." Bohm went well above my intellectual head, but left me with an intriguing new sense of underlying mystery.

Seth came to my attention later. After Jack Sarfatti, Bob Toben, and I published our popular book *Space-Time and Beyond*, a reader told us that we had explained, in the terminology of modern physics, the very same things that Seth talked about. We had not intended to go into the subject in any real depth; we simply wanted to sketch the relationship between science and spirit in a very light way. Mr. Friedman has developed those sketches, has looked deeply into these comparisons, and has written with skill and insight. Although we had certainly placed our feet on the bridge in our attempt to "explain the unexplainable," it was left to Norman Friedman to cross over.

The essential element in all of this, I would say, is that all viewpoints of understanding our experience of the Universe rest on the simultaneous existence of a deeper level of reality out of which the duality of the physical and mental aspects emerge.

The book is divided into two parts. The first part clearly explains the unity that underlies the three separate perspectives of deeper reality: David Bohm's concepts of how physics leads to this underlying element, Ken Wilber's description of how mystics experience it, and Seth's discussion of the hidden reality from the unique perspective of one who lives there. After this introduction to the strange landscape of the new physics world, Part Two presents a range of topics expressed in metaphors taken from modern science, including not only physics but mind science and dream research.

A particularly apt image for the view we will gain from Mr. Friedman's book comes from Edwin A. Abbott's *Flatland*. In the words of Mr. A. Square, when he was taken by a strange sphere from his flat world of two dimensions into the bewildering reality of three-dimensional space:

> An unspeakable horror seized me. There was a darkness; then a dizzying sensation of sight that was not like seeing; I saw . . . Space that was not Space: I was myself, and not myself. . . . I shrieked . . . , "Either this is madness or it is Hell." "It is neither," replied the voice of the sphere, "it is Knowledge; it is Three Dimensions: open your eye once again and try to look steadily."

> I looked, and, behold, a new world!

For many of us, the new landscape we see from the bridge built by Norman Friedman may seem as strange as the third dimension did to Mr. A. Square. We have lived our lives assuming that the dimensions marking the spirit lie well beyond those marking the physical landscape. The proposition that the two are identical is bewildering to say the least. But this book encourages us to look boldly, and with Mr. Friedman as our able guide, we need only to open our eyes to "behold a new world."

Fred Alan Wolf, Ph.D.

Author of *Taking the Quantum Leap, Parallel Universes, The Eagle's Quest, The Dreaming Universe*, and other books.

La Conner, Washington

Preface

*B*efore we begin this journey together, two confessions are in order.

First, I am a passionate fan of contemporary physics. For many years I have followed the philosophical convulsions resulting from quantum theory and relativity theory. But I am not a practicing physicist, I am a spectator. My view of the game — the ongoing quest to explain reality — is from the bleachers rather than from the dugout and the field. In some sense I have played in the minor leagues (with a bachelor's degree in physics and a master's degree in engineering), and that allows me to follow most of the plays fairly closely. Although my mathematical background lacks the depth and breadth to allow me to hit a metaphorical 90-mile-an-hour physics fastball, I understand the principles involved in such a feat.

The view from the bleachers has some advantages. I am unencumbered by the demands of daily practice and regular competition. My perspective is broader than it would be if I were engaged in the action of the game. I am free from restrictions that go along with being a member of the team. As a fan, I can applaud certain plays and players and boo others with impunity. If my analysis of a game differs from that of the manager or players, it is a matter of no great consequence.

But while asserting that there is something special about my view from the bleachers, I do so with a certain amount of humility. I am always aware of my dependence on the players, and I respect their talents as professionals. Accordingly, in the following pages I make abundant use of the knowledge, comments, and ideas of scientists and philosophers.

The second confession is that my interest in this game has intensified into a burning curiosity, which has led me far from the familiar playing fields of scientific investigation. In short, this book includes ideas not only from physics (primarily the work of David Bohm) but of mystics, represented by Ken Wilber's treatment of the Perennial Philosophy, as well as — hold on to your seats — the channeled spirit entity known as Seth. Although the methods, concepts, and language of these three sources vary markedly, parallels in their descriptions of reality are striking indeed.

No doubt you are wondering what a nice physics fan is doing in the company of mystics and mediums. In explaining, let me begin by saying that at least I am not alone in this. In recent decades, many books have compared the experiences recounted by mystics to the philosophical implications of modern physics. Best known of these, perhaps, is Fritjof Capra's *The Tao Of Physics*. Capra is uniquely qualified to comment on these connections because of his research in high-energy physics and his mastery of meditation techniques. His insights are necessarily general because of the differences in approach between the mystic and the physicist, and because the mystical experience is by nature ineffable, since its aim is alignment with the whole.

Physicists, on the other hand, analyze reality as separate, describable parts. While they may be aware of the interconnections of the entire universe, their methodology involves examination of particulars. Still, the *interpretation* of physical particulars can convey a more encompassing reality. Physicist David Bohm presents us with such a view. Bohm's ideas and those of Ken Wilber have been compared in several publications, including interviews in *The Holographic Paradigm And Other Paradoxes*. Because this material is available, and because my main interest lies in the particulars of physics, the Perennial Philosophy is considered only briefly here.

My introduction to the more esoteric figure of Seth was through an article by quantum physicist Nick Herbert, which included the following statement:

Jane Roberts, in her Seth books, describes, as an aspect of human personality, a world of "probable realities" in which opposites coexist in a manner similar to Heisenberg's potentia. It is too early to say whether notions like these are mere casual analogies or indications that we are on the threshold of a new sensual physics.[1]

At the time, I had read very little about the paranormal and had never heard of Jane Roberts, but I immediately bought a Seth book, *The Individual And The Nature Of Mass Events*. In it I found discussion of a "framework" that sounded similar to Heisenberg's world of potentia and the ghost field of quantum physics. I read other books by Roberts/Seth and began culling those portions related to physics. The more I read, the more I saw that the descriptions of reality by physicists and by mystics were *bridged* by those of Seth. Such connections are mind-boggling, but as physicist Freeman Dyson has said, "For any speculation which does not at first glance look crazy, there is no hope."[2]

The source of the Seth material will not concern us here. If Seth is a deception by Roberts, it is a remarkable one, for it would require a grasp of science and philosophy that would be extremely unusual considering her background as a poet and novelist. On the other hand, if the Seth material originated at some unknown level of Roberts's unconscious mind, then that level must be a repository of knowledge far beyond our normal awareness.[3] Regardless of the "true" identity of Seth, the ideas expressed in this manner are invaluable in clarifying certain relationships between science and mysticism.

Although this book is intended for the general reader, some parts will be difficult without a background in science. (Many of the chapter notes contain information provided for readers whose understanding of scientific material is fairly advanced.) For those readers who find certain sections too demanding, my suggestion is to persist, skimming or even skipping the hard parts until you reach a more comprehensible section. It is my experience that allowing yourself to provisionally accept some of these ideas, even if you don't entirely follow the reasoning behind them, may open the way to a general and often deeper understanding later on.

Now that our confessions are made and our paths roughly mapped out, let us begin our journey.

Introduction

A religion contradicting science and a science contradicting religion are equally false.

P. D. Ouspensky

We all want to know that the universe is a harmonious place in which we can live in comfort and peace. Some of us may sense that intuitively. But for those of us who want scientific evidence, answering the question, "What's it all about?" would seem to require comprehension of vast spheres of knowledge beyond our grasp. Metaphors reduce that requirement to something more attainable. That is, an image we can empathize with condenses the vastness of the universe into a meaningful symbol and satisfies much of our need to know.

My purpose in this book is to present images of reality that are both illuminated by spirit and grounded in science. In this pursuit, I have been inspired by the following words of physicist Mendel Sachs:

> It behooves those who seek truth to study the abstract features of the truths of as many disciplines as possible, in order to determine which of the ideas of each of them correspond and which do not, with the notion that those ideas that do recur in a varied range of domains of knowledge are more likely to be true than those that do not. Thus the seemingly invariant truths are the ones that should be pursued further, as significant investigations toward our future understanding of the real world.[1]

Following this advice to truth-seekers, I have examined three radically different frames of reference: aspects of theoretical physics, mysticism, and the paranormal. I have found in these three sources enough similarities to construct what I believe to be a coherent image of reality that could open the door to a profound new understanding of the universe.

The first of these is the work of David Bohm, a theoretical physicist. His concept of reality is indirect, arising from the logical consequences of the algorithms of quantum theory and relativity theory.

The second is Ken Wilber, a modern authority on the Perennial Philosophy. His overview is based on the accounts of mystics who, through altered states of mind, have directly experienced a mysterious domain beyond our ordinary perceptions. Wilber has distilled from this material (much as I have attempted here, albeit with more disparate sources) a common conceptual framework that underlies all experience and gives a spiritual dimension to that fragment of it that we call everyday reality.

Our third source is "channeled" through a medium. He is Seth, a discarnate personality whose prolific views on the nature of reality were communicated through Jane Roberts. Although this notion is guaranteed to elicit strong emotional reactions along the range from skepticism to scorn, the ideas presented by this "entity" are not only intriguing but intellectually challenging. As we shall see — and in considerable detail, for it constitutes the main thrust of this book — Seth's descriptions are remarkably consistent with the physical theories of David Bohm.

The ideas of Bohm, Wilber, and Seth, presented through the vastly different means of reason, revelation, and channeled communication, are the subject of Part One. Considered together, they at once broaden our perspective and act as a lens that refocuses our examination of reality. In Part Two, with this lens before our eyes, we examine six of the most thought-provoking issues of our time, which leads us into perplexing but fascinating territory of the fundamental nature of the universe.

CONSCIOUSNESS: A PARADOX

Like all scientific and philosophical constructs, what I propose in this book is tentative, incomplete, and subject to change. This is especially true of a treatment as broad as the one presented here, encompassing as it does such diverse aspects of our knowledge territory. Let us begin boldly, then, by stating that at the center of this whole discussion is consciousness, which is not easily definable, for reasons that will become clear in Chapter 1.

Consciousness heretofore had no place in the equations of physics. In the classical model of the universe, irreducible balls of matter bounced about in three-dimensional space, obeying fixed laws of motion. This universe was both objective (independent of an observer) and determined (predictable). Modern physics offers a much less comforting picture. Quantum theory rests on a bed of indeterminacy at the particle level, and, while the predictive power of quantum theory is awesome, its philosophical underpinnings are vague. Reality cannot be said to exist in any fixed, solid way, and even the physical nature of matter is questionable. How, then, do we account for our perception of such things as buildings and trees? With this Zen-koan-like proposition: that the building is real, but does not have existence until it is observed. In the words of physicist John Wheeler,

> No elementary phenomenon is a phenomenon until it is a registered (observed) phenomenon. . . . Useful as it is under everyday circumstances to say that the world exists "out there" independent of us, that view can no longer be upheld.[2]

The paradox is this: we need particles of matter to make up the objects of our everyday world (including us), and we need an object in that very everyday world (us) to define and observe those particles. Observation implies consciousness. Most physicists resist this implication, but at least some would agree that any construct that purports to describe reality in terms of contemporary physics clearly must include a role for consciousness.

In the late nineteenth century, the theoretical structure of physics appeared to be virtually complete, raising the distinct possibility of eventually explaining everything. Newton's laws and Maxwell's equations provided a mathematical foundation for the universe, a universe considered to be thoroughly predictable. Since every cause had an effect, and every effect had a cause, if one knew the initial conditions of a state, its past and future could be ascertained. Physicist Banesh Hoffman, a former member of the Institute for Advanced Study at Princeton, described in this way the air of complacency that permeated nineteenth-century physics:

> Had it not now reduced the workings of the universe to precise mathematical law? Had it not shown that the universe must pursue its appointed course through all eternity, the motions of its parts strictly determined according to immutable patterns of exquisite mathematical elegance? Had it not shown that each individual particle of matter, every tiny ripple of radiation, and every tremor of ethereal tension must fulfill to the last jot and tittle the sublime laws which man and his mathematics had at last made plain? Here indeed was reason to be proud. The mighty universe was controlled by known equations, its every motion theoretically predictable, its every action proceeding majestically by known laws from cause to effect.[3]

Physicists of the time felt confident that they had deciphered God's plan.

This view applied to the entire universe, organic life and sentient beings included. An attribute of life such as consciousness, which arises from a preferred arrangement of elementary particles, was considered an epiphenomenon of the underlying structure, and, in principle, could be explained fully by the basic constituents and the forces between them. As Voltaire said,

> It would be very singular that all nature, all the planets, should obey eternal laws, and that there should be a little animal, five feet high, who, in contempt of these laws, could act as he pleased, solely according to his caprice.[4]

An ongoing debate arose among some scientists and philosophers over how consciousness originates from matter. (That the question might be reversed — that is, how does consciousness produce matter? — has been and still is largely inconceivable.)

In expressing the view of his time that physics was essentially complete, a leading theoretical physicist, William Thompson (Lord Kelvin), noted only two exceptions: the failure to detect the ether and the failure to understand black-body radiation (a body that absorbs all radiation falling on it). Lord Kelvin's confidence not withstanding, these minor matters not only refused to disappear, but by the turn of the century took on major importance. The black-body radiation problem led to quantum mechanics (dealing with subatomic particles), and the negative results of the Michaelson-Morley experiment (failure to detect the ether) led to Einstein's theory of relativity — the foundations of modern physics. Classical physics was thereby reduced to a special case of these new, more encompassing principles.

In relativity theory, the particle described in classical physics is no longer a "thing" but a vortexlike disturbance in a continual field. Consciousness was introduced into physics in a sense because relativity theory requires that the frame of reference of the observer be taken into account. In quantum theory the introduction of consciousness is even more basic. Physicist Fred Alan Wolf puts it this way:

> Classical physics holds that there is a real world out there, acting independently of human consciousness. Consciousness, in this view, is to be constructed from real objects, such as neurons and molecules. It is a byproduct of the material causes which produce the many physical effects observed.
>
> Quantum physics indicates that this theory cannot be true — the effects of observation "couple" or enter into the real world whether we want them to or not. The choices made by an observer alter, in an unpredictable manner, the real physical events. Consciousness is deeply and inextricably involved in this picture, not a byproduct of materiality.[5]

Exactly how the relationship between consciousness and matter occurs is still an open question. In fact, it will be central to our discussion in the chapters to follow.

THE MIND IN PHYSICS AND BIOLOGY

Harold J. Morowitz, a molecular biophysicist, wrote a perceptive article in 1980 illustrating the changes in the worldview emerging from the new physics and its effect on the psychological and biological sciences. He said:

> Something peculiar has been going on in science for the past 100 years or so. Many researchers are unaware of it, and others won't admit it even to their own colleagues. But there is a strangeness in the air.[6]

Morowitz then described an interesting situation. Biologists originally believed that the human mind occupies a special position in the scheme of things. Influenced by the rampant successes of nineteenth-century physics, however, they shifted toward a more mechanistic orientation. But around 1900, physics changed direction in response to the quantum and relativity theories, and the mind began to assume an essential role in physical events. (The participation of an observer — or consciousness — was codified by the Copenhagen interpretation in the 1920s; see pp. 31-32.) News of this fundamental change apparently never reached the biologists, for they rushed headlong into materialism, while the physicists moved in the opposite direction. Morowitz used the metaphor of two fast-moving trains going in different directions, each unaware of the activity on the other track.

For the past eighty-five years biological and psychological sciences have largely relied on reductionist methods, that is, explaining phenomena at a higher and more complex level by phenomena at a lower and more basic level. This approach is affirmed by Carl Sagan in *The Dragons of Eden*: "My fundamental premise about the brain is that its workings — what we sometimes call 'mind' — are a consequence of its anatomy and physiology and nothing more."[7] Morowitz comments on Sagan's book:

> As a further demonstration of this trend of thought, we note that his glossary does not contain the words "mind," "consciousness," "perception," "awareness," or "thought," but rather deals with entries such as "synapse," "lobotomy," "proteins," and "electrodes."[8]

Thus, biology has busily pushed the mind out one door of the house of science, little realizing that it was reentering through the door of physics. Morowitz sums it up cogently:

> We are now in a position to integrate the perspectives of three large fields: psychology, biology, and physics. . . . [resulting in] a picture of the whole that is quite unexpected.

> First, the human mind, including consciousness and reflective thought, can be explained by activities of the central nervous system, which, in turn, can be reduced to the biological structure and function of that physiological system. Second, biological phenomena at all levels can be totally understood in terms of atomic physics,

that is, through the action and interaction of the component atoms of carbon, nitrogen, oxygen, and so forth. Third and last, atomic physics, which is now understood most fully by means of quantum mechanics, must be formulated with the mind as a primitive component of the system.

We have thus, in separate steps, gone around an epistemological circle — from the mind, back to the mind. The results of this chain of reasoning will probably lend more aid and comfort to Eastern mystics than to neurophysiologists and molecular biologists: nevertheless, the closed loop follows from a straightforward combination of the explanatory processes of recognized experts in the three separate sciences. Since individuals seldom work with more than one of these paradigms, the general problem has received little attention.[9]

This is confirmed by the following comments. Werner Heisenberg, father of quantum theory:

Some scientists [have been] inclined to think that the psychological phenomena could ultimately be explained on the basis of physics and chemistry of the brain. From the quantum-theoretical point of view there is no reason for such an assumption.[10]

Physicist Paul Davies, a leading popularizer of modern physics:

It is often said that physicists invented the mechanistic-reductionist philosophy, taught it to the biologists, and then abandoned it themselves. It cannot be denied that modern physics has a strongly holistic, even teleological flavour, and that this is due in large part to the influence of the quantum theory.[11]

THE OBSERVER IN PHYSICS

The fact that an observer — which in itself implies consciousness — is an established and necessary ingredient in modern physics has profound implications. When scientists describe the world as made up of particles, they must of necessity include themselves in the construction. By definition, then, the scientist is a part of the universe, and this part of the universe is observing itself. To accomplish this, the universe must be divided into an observer and that which is observed. What the observer sees is only that portion of the universe that is being observed, which does not include the observer. A new level with a wider field of view must be postulated if we are to include both the observer and that which is observed. Thus, we are forced to consider a hierarchy of consciousness. A hierarchy is an essential metaphor for visualizing both the observer and that which is observed. (Such a hierarchy is a simplified version of the Gödel incompleteness theorem, which can be construed as stating that if you have a finite theory of the world, there will always be certain truths that will not be provable by the theory; see discussion p. 180.)

A hierarchy is also seen in quantum theory. One of the basic postulates of quantum theory states that an initial system can potentially develop into a number of states, each with a given probability of occurrence. We know, though, that only one probability can actually take place. According to the accepted interpretation of quantum theory, the actual or "real" state of these probabilities is specified by an observation; that is, the "real" state is brought into reality by an observer. The observer thus becomes a creator and gives the system its form. Without the observer, the system is in a state of potential, waiting to come into existence.

This approach has proved fruitful for closed systems with the physicist observing from outside. But when one contemplates the grouping of probabilities for the entire universe — including the body of the observer — then the universe can come into existence only by the action of an observer outside the universe. Even if one were to accept an observer outside the universe, that observer would also have to be brought into reality by still another observer. Again, we are faced with an infinite hierarchy of observers.

According to quantum theory, mind gives form to *potentia* (Heisenberg's term), which then exhibits the property of matter. We have come full circle: matter appears to be an epiphenomenon of mind. This does not mean, however, that the electron is dependent on the *human* mind. Rather, if the electron is seen in some sense as being "alive," then it can have its own equivalent of observer and observed, its own form of consciousness.

If we no longer see *matter* as primary, but see *mind* as primary, then the push-pull laws of the mechanistic worldview are not adequate. Instead we might say that an "alive" particle responds to information, and force fields (e.g., electromagnetic and gravitational) might be viewed as information fields. Seth comments on the primacy of consciousness:

Science has, unfortunately, bound up the minds of its own most original thinkers, for they dare not stray from certain scientific principles. All energy contains consciousness. That one sentence is *basically* scientific heresy, and in many circles, it is religious heresy as well. A recognition of that simple sentence would indeed change your world. . . .[12]

A somewhat similar view is expressed by physicist Freeman Dyson:

It's one of the joys of physics that matter isn't just inert stuff. In the Nineteenth Century one thought of matter as just chunks of stuff which you could push and pull around, but they didn't do anything. Quantum mechanics makes matter even in the smallest pieces into an active agent, and I think that is something very fundamental. Every particle in the universe is an active agent making choices between random processes.[13]

CASTING A FINER NET

It is often said that most phenomena can be described with mechanical models — that the Newtonian laws hold for large-scale events, and that it is only at the far reaches of existence (the subatomic world, or particles moving at nearly the speed of light) that we see the peculiar effects of quantum and relativity theories. But this is not necessarily so. Events that deviate from Newtonian laws may occur in our everyday world without being observed simply because we are not looking. The following parable, attributed to Sir Arthur Eddington, the distinguished astrophysicist, raises this possibility using a particularly apt image.

In a seaside village, a fisherman with a rather scientific bent proposed as a law of the sea that all fish are longer than one inch. But he failed to realize that the nets used in the village were all of a one-inch mesh. Are we filtering physical reality? Can we catch consciousness with the nets we are using?[14]

In a remarkably similar statement, Seth said:

Science itself must change, as it discovers that its net of evidence is equipped only to catch certain kinds of fish, and that it is constructed of webs of assumptions that can only hold certain varieties of reality, while others escape its net entirely.[15]

Do anomalies exist on the macro level that we are simply not noticing or are allowing to slip through our scientific nets? Certain events, casually assigned to the paranormal and thereby dismissed, may contain information that would clarify persistent problems in our evolving worldview. Philosopher Huston Smith imagines the Perennial Philosophy addressing science and noting, "You are right in what you affirm. Only what you deny needs rethinking."[16]

Certainly many insupportable ideas have been rightfully ignored, but one can speculate that useful information may have been overlooked as well. This is by no means confined to the science of our time. In the 1600s, the great Galileo Galilei (himself persecuted for heretical ideas) wrote the following attack on Johannes Kepler's idea that the moon affects the tides:

> Everything that has been said before and imagined by other people [concerning the origin of tides] is in my opinion complete nonsense. Among authorities who have theorized about the remarkable set of phenomena, I am most shocked by Kepler. He was a man of exceptional genius, he was sharp, he had a grasp of terrestrial movement, but he went on to take the bit between his teeth and get interested in a supposed action of the moon on water, and other "paranormal" phenomena — a lot of childish nonsense.[17]

Perhaps, without being aware of it, science has narrowed its field of vision and enlarged the holes of its net. Of course, the quest to understand reality can never be completed. Some information will always be lacking; some ambiguity will always remain. Although there may be many brilliant insights, a total explanation is not and cannot be the goal. Rather, we try to unfold a bit more, expand our vision, and extend our understanding. Perhaps we can create nets with a finer mesh so that fewer fish escape our notice — including that slippery one Eddington mentions, consciousness. The main aim of this book, we might say, is to cast a finer net.

PART ONE

Three Perspectives, One Reality

1

David Bohm:

A Physics Perspective

Recent decades have taught us that physics is a magic window. It shows us the illusion that lies behind reality — and the reality that lies behind illusion. Its scope is immensely greater than we once realized. We are no longer satisfied with insights only into particles, or fields of force, or geometry, or even space and time. Today we demand of physics some understanding of existence itself.

John Wheeler

*M*odern physics has not given rise to a commonly acknowledged metaphor for reality. There is no single conceptual framework that fits seemingly unrelated phenomena into a meaningful pattern, provides an image that physicists use when they attempt to construct a model of their equations, and serves as a guide to research. Although there are contenders for such a framework, none has been generally accepted. This seems strange, since the basic theories of modern physics (quantum theory and relativity theory) have been around for over sixty years. However, the mathematical formalisms on which these theories are based are not easily represented by visual symbols and thus do not lend themselves to metaphor. Physicists, on the whole, are not disturbed by this. If pressed, they would probably endorse the Copenhagen interpretation set forth by Danish physicist Niels Bohr that there is no objective reality on the quantum (subatomic) level.

The Copenhagen interpretation is startling: the basic constituents of matter are not "things" at all! Quantum particles are *not* objective particles like billiard balls or baseballs;[1] nature at the quantum level is detectable only through meter readings or computer readouts that relate to esoteric mathematical formulations. According to this view, an electron cannot be said to have a position or momentum (or any dynamic attribute) until that property is measured. (In essence, we "look" at quantum particles with measuring devices such as bubble chambers.) Not only do we not know the position and momentum of an electron until we measure them, but in between measurements, such terms have no meaning. To state it another way, the electron *unobserved* is not real. In the words of Werner Heisenberg, "If we want to describe what happens in an atomic event, we have to realize that the word 'happens' can apply only to the observation, not to the state of affairs between two observations."[2] When either the position or momentum of an electron is measured, the other, unobserved property becomes uncertain. Thus the trajectory of an electron can be predicted only as a probability.

The Copenhagen interpretation effectively rejects the model of Newtonian mechanics. Lord Kelvin said that he could not *really* understand the electromagnetic theory of light because it was not picturable. Imagine his consternation if he had to deal with a modern-day quantum particle! If the universe is not like a machine or a steam engine, as the classical physicists imagined, what is it like? According to the Copenhagen interpretation, we do not know what reality is, and, more important, we can never know. Most physicists feel that there is no point in worrying about a philosophical basis for reality because their purpose in physics is to make models or equations that help in correlating observations. From the physicist's point of view, if quantum mechanics merely describes what can be observed in a measuring device, so be it; there is no such thing as a "real" world out there anyway.

Physicist Murray Gell-Mann called quantum mechanics "that mysterious, confusing discipline which none of us really understands but which we know how to use."[3] Since physics describes the "how" of things, not the "why," it is easy to ignore the philosophical problems raised by the Copenhagen interpretation. Philosopher and scientist Bertrand Russell suggested an analogy that illuminates this point. He asked us to compare music played by an orchestra with the same piece as it is printed in the score. Imagine a person deaf from birth who has been raised and taught by musically trained parents. Such a person could conceivably understand all the abstract characteristics of the music, even "hear" it from reading the score, but its experiential value would be completely unimaginable. To Russell, the physicist's knowledge of nature is similar to that of the person with musical training who has never known sound: correct, but missing the essence — the music.

THE PHYSICS METAPHOR: A SECOND LEVEL

A small number of physicists are decidely dissatisfied with the uncertainty of a probabilistic model of reality. Those dissatisfied few "reality seekers" have proposed various solutions in their search for a mathematical construct that will bring determinism back into physics. Einstein himself sought an answer, but although he could not accept the indeterminism of quantum theory, he could not find an alternative. He felt that "underneath"[4] the statistical nature of the new quantum theory, there had to be more. The quest of Einstein (and others) for a complete theory became known as the search for "hidden variables," the assumption being that there are variables not yet known that will explain the probabilistic aspect of events. If such variables were discovered, present quantum theory, though still applicable in its domain, would turn out to be a special case of the newer, more complete theory, and we might have a more understandable philosophical underpinning for modern physics.[5]

Heisenberg, in an extension of the Copenhagen view, described the hidden world beneath our phenomenal world as one of *potentia*, a term first used by Aristotle. All possible events exist on this level in potential form. That is, particles do not have real existence but have *tendencies* to exist. In this nether world, particles can be in contradictory states (e.g., occupying two different positions at the same time); it is our act of measurement that brings one or the other state into reality. According to physicist Hugh Everett and his followers, however, all these probabilities and contradictory states do have an existence. Since it is obvious that our macroscopic world cannot accommodate contradictory states, all the unused states must go elsewhere to find actuality. For the "elsewhere," Everett postulates parallel worlds that exist alongside each other, but without contact since they occupy parallel planes.

Concerning the case for parallel worlds, physicist Fred Alan Wolf has this to say:

> The science-fictionlike idea that our universe is not alone — that there exist in some mysterious manner alongside of ours (and this needs some explaining) other universes — is the latest concept brought forward by the new physicists in their attempt to unify our knowledge. Without the existence of these other worlds, these gaps of knowledge brought into light by the discoveries of the new physics would remain unbridgeable — incapable of being solved by previous thinking.[6]

Another physicist, Eugene Wigner, states outright that consciousness must be introduced into the laws of physics. In his view, sentient consciousness makes the choice between the various contradictory states.

The important point is that some physicists who address the nature of reality are forced to postulate a world underneath, alongside, or perpendicular to our more familiar world. That is, there appear to be hidden dimensions to reality. Even the Copenhagen interpretation involves a second level, in that to bring the quantum particle into "reality" — the phenomenal world, or first level — a measuring device is necessary. The phenomenal world arises from choices made from the second or hidden level. How these choices are made is central to the ideas discussed in this book.

According to these interpretations, the basic operations of our three-dimensional world are manifestations of something going on outside our perceived space. The events of our world require an underlying dimension or process. This hidden arena is not definable in mechanistic terms; it is a vast conglomeration of tendencies, contradictions, and probabilities. Such a world bears little resemblance to our everyday existence.

Physicist John Stewart Bell proved that if the experimental results of quantum theory are correct, any hidden variable theory must be "nonlocal."[7] Nonlocal means that if two particles interact and then move apart, they continue to influence each other instantaneously —

faster than the speed of light — in spite of the fact that, according to special relativity theory, signals in our three-dimensional world cannot exceed the speed of light.[8] Bell says that, although it pains him even to consider it, his theorem shows that "there must be something happening faster than light . . . . The theorem certainly implies that Einstein's concept of space and time, neatly divided up into separate regions of light velocity, is not tenable."[9] The theorists who want to substantiate a deterministic view, then, will have to look elsewhere than our familiar three-dimensional world.

Thus, modern physics, in attempting to reconcile the special theory of relativity and Bell's theorem, faces a profound philosophical dilemma. On the one hand, Bell's theorem states that reality must be nonlocal, since in a local reality information can be transferred at speeds no greater than the speed of light, and such speeds are insufficient to explain the quantum facts. Note that Bell's theorem depends on quantum *facts*, not quantum *theory*. Quantum facts are established by experiment. Since Bell's work is based on established facts, his theorem cannot easily be dismissed. On the other hand, special relativity theory states that the speed of light cannot be exceeded. Bell has commented on the problem, "I think it's a deep dilemma, and the resolution of it will not be trivial; it will require a substantial change in the way we look at things."[10] To accommodate both special relativity theory and his own nonlocal view of reality, Bell postulates the existence of another (or "deeper") level than our universe. The instantaneous interconnectedness of quantum particles must be accomplished *outside* our three-dimensional universe, in some sort of extradimensional space. Again, the description of the universe requires at least a second level.

As a result of this dilemma, David Bohm has suggested a field that does not inhabit space-time, a transcendental realm in which all places and all times are merged. If this seems strange, remember that any solution will have to be literally out of this world if nonlocal reality and special relativity theory are both to be satisfied. Bell's theorem suggests that three-dimensional space and a fourth dimension of time are insufficient to completely explain particle interaction. An underlying reality with a no-time attribute is required. Even Einstein once observed that "time and space are modes by which we think and not conditions in which we live."[11] Bell's theorem tells us that this indeed may be true.[12]

BOHM'S IDEAS: AN OVERVIEW

David Bohm (1917–1992) grew up in Wilkes-Barre, Pennsylvania, was educated at Pennsylvania State College, and received his doctorate from the University of California at Berkeley. After working briefly for the Manhattan Project in 1943, he accepted an assistant professorship at Princeton. In 1951, because of his association in graduate school with J. Robert Oppenheimer, Bohm was called before the Committee on Un-American Activities of the House of Representatives. Like many others of his time, Bohm refused to testify, citing the Fifth Amendment. When his contract at Princeton expired in 1951, he was unable to find a position in the United States. The infamous McCarthy period had produced another casualty in the American scientific community. During the next ten years, Bohm held professorships at universities in Sao Paulo and Haifa and a research fellowship at Bristol University. In 1961 he was named Professor of Theoretical Physics at Birkbeck College in London. He retired from that position in 1983, but continued his research until his death.

During his years at Princeton, Bohm wrote what many physicists consider the classic text on quantum theory. In general, physics books are not only dry but liberally laced with complicated mathematical formulae. Wolf, who read Bohm's *Quantum Theory* in a graduate course, found it to be an exception.

> It had more words than formulae. It dealt with questions concerning topics that were apparently unrelated to physics. "The Indivisible Unity of the World," "The Need for a Nonmechanical Description" of nature, "The Uncertainty Principle and Certain Aspects of Our Thought Processes," and "The Paradox of Einstein, Rosen, and Podolsky" were some of the topics that Bohm considered and that were to have a great influence on my own thinking.[13]

Bohm's book was consistent with the Copenhagen construct of reality — an attempt, in fact, to understand quantum theory from Niels Bohr's point of view. Soon after its publication, however, Bohm began having second thoughts. His main concern was with the concept of motion. The Copenhagen interpretation saw motion as a series of discrete events designated by a collapsing wave function, in contrast to the classical view of continuous progression along a trajectory. The notion of a collapsing wave function is essential in contemporary physics but is difficult to visualize. A short digression into its historical background will help us understand its significance.

Particle and Wave

During the early part of the twentieth century, the physics community became aware that electromagnetic radiation, so beautifully described by Maxwell as a wave field, also displayed particle characteristics. This development was disconcerting for those physicists who felt a need for a mechanical model of reality. The idea that electromagnetic radiation could have both wave *and* particle characteristics was difficult to entertain,[14] but the experimental work published by physicist Arthur Compton in 1922 firmly established the concept. Compton found that an x-ray (normally thought of as a wave) acted just like a photon (an electromagnetic particle). Whether electromagnetic radiation should be denoted as a particle or as a wave depended on what experiment was performed — what question the physicist put to nature. (It should be emphasized that making this verbal distinction is only a convenience since both particle and wave are aspects of the same thing.)

If a wave could appear as a particle, it was only natural to ask if a particle could appear as a wave. Two years after Compton published the results of his experiment, physicist Louis de Broglie discovered one of the most astonishing facts of quantum theory. He worked out mathematical formulations for connecting the particulate and wave properties of *all* matter, micro or macro, not just electromagnetic radiation. This discovery placed the elementary particle (and therefore all matter) in a whole new light, since it could no longer be visualized as similar to a small billiard ball. You can sense the excitement in de Broglie's account of his sudden realization that "Einstein's wave-particle dualism was an absolutely general phenomenon extending to all physical nature."[15] However, because the mass of a body in macro reality is so large, the wavelength is too short to be detectable, so we are not generally aware of wavelength as applied to everyday items.

On the particle level, the dual properties are more apparent, and in fact, in 1927, physicists Clinton J. Davisson and Lester H. Germer demonstrated the wave properties of electrons in the laboratory.

Most of the older physicists of the 1920s had been trained in the classical tradition. To deal with this new conception of waves, it was natural for them to look for a wave equation. Erwin Schrödinger proposed such an equation in 1925. At first it was assumed that Schrödinger's wave equation would be similar to that describing an electromagnetic wave, a wave considered "real" (i.e., it carries energy).[16] But the new quantum wave equation, which was identified with de Broglie's waves, turned out to be quite different. The quantum wave does not carry energy as we know it. It could not exert a force on the mass involved, or even be measured. The reason for this characteristic is that the wave function, a solution of Schrödinger's equation, is a complex function[17] and therefore can never be observed directly.

Normally, in classical physics, the energy of a wave is proportional to the amplitude squared. In the case of Schrödinger's wave, which has no energy, what does the amplitude squared mean? Physicist Max Born interpreted the amplitude squared as the probability of finding a particle at that location if a detector is placed there, thus making the quantum wave not "real" in our three-dimensional world.[18] Obviously, this was not a wave that a classical physicist could be comfortable with. Einstein went so far as to call the quantum wave field a "ghost field." Some physicists, such as Henry Margenau, saw the wave function as the "true" reality. In spite of the lack of clear understanding of its meaning, almost all physicists use Schrödinger's wave equation as a calculating device, and, to say the least, it works extremely well.

To conclude this digression, de Broglie discovered that *all* matter, whether an electron or the entire universe, has wavelike aspects. There is, in principle, a wave function associated with all configurations of matter. These wave functions are solutions of Schrödinger's wave equation. We never "see" this wave field, but we know its effects through measurements. In some extraordinary way, the field is not in our three-dimensional universe, though its consequences are. In the Copenhagen view, the quantum field specifies the probability of finding a particle at a given position in space if one happens to be looking;

the particle is a manifestation of the field when observed. When we look at it, it is a particle. When we are not looking, it is a wave, but a wave that is not real.

This effect of looking and not looking makes observation a key concept in modern physics. The entire wave appears to *collapse* down to a particle instantaneously upon being observed. The wave function describes conditions before matter makes its appearance; we never see the wave until it collapses into matter. In essence, for every formation of matter there is a corresponding wave function, which contains all its probabilities of activity. But the wave function is essentially *passive*; mathematically speaking, it is linear. It cannot stimulate action from within itself. It requires *an agent* to make a choice among its probabilities for the three-dimensional world to be formed.

Some physicists, such as Wigner, identify this agent as sentient consciousness. In other words, human consciousness is what stimulates action in the otherwise passive wave function. Bohm feels that there is such an agent within all matter that performs this function. Perhaps because of his conversations with Einstein, Bohm became convinced that a hidden variable theory was possible and that quantum theory was not the last word. In terms of the Copenhagen interpretation of quantum mechanics, motion was a series of observations (collapsing wave functions) dependent on the physicist being present to do the observing. To Bohm this description was not adequate.

The Hamilton-Jacobi Theory of Motion

In 1952, Bohm published a paper titled "A Suggested Interpretation of the Quantum Theory in Terms of Hidden Variables" in *Physical Review*. This interpretation grew out of similarities he had noticed between the Schrödinger wave equation and the equations formulated in the late 1860s by William Hamilton and Carl Jacobi. Hamilton and Jacobi developed a concept of the mechanics of moving bodies based on waves rather than particles. In this unusual departure from Newtonian mechanics, they stated that the particle and its movements are produced by an underlying wave motion; that is, the particle can be seen as moving perpendicular to the wave.

Bohm uses the image of a body of water to clarify this concept. Suppose a group of waves combine in one area of the water. This area exhibits a disturbance called a wave packet, while outside this area, the disturbance is negligible. The collection of wavelets appears as a particle. In addition, the packet maintains itself by constantly folding in and out of the body of water. (Bohm uses the terms enfolding and unfolding.) The packet, or the particle, moves along with what is known as a group velocity. The motion is determined by the ebb and flow of the entire wave rather than by local forces acting on the particle at each point of its trajectory. In other words, the packet is totally controlled by the ground (body of water) from which it was created. While it is feasible to treat this collection of wavelets as independent units in some selected cases, it really is part and parcel of the whole body of water. What we see are patterns, sustained by an unfolding and enfolding process and then retreating into their original home, the main body of water.

The Hamilton-Jacobi view of motion allowed for a wavelike conception that was congruent with the particle conception. Bohm contends that most scientists completely missed the significance of the Hamilton-Jacobi approach. Here was a theory that produced the same results as Newton's theory, yet treated matter as having the characteristics of both waves and particles. Thus, Hamilton and Jacobi provided a metaphor for quantum theory many years ahead of its time. As Bohm points out:

> Such a metaphor that connects the essence of the two theories would have, in a certain sense, anticipated the modern quantum-mechanical notion of wave-particle duality. That is, that the same entity (e.g., an electron) behaves under one set of circumstances as a wave, and in another set of circumstances as a particle. A more detailed discussion and development of this metaphor could have led in the mid-nineteenth century to the general outlines of the modern quantum theory, almost without any further experimental clues at all.[19]

However, scientists were so caught up in the mathematics of the Hamilton-Jacobi theory and so trapped by the Newtonian models that the quantum aspect was completely overlooked.

The Hamilton-Jacobi interpretation allowed Bohm to understand motion in a more satisfying way than did previous conceptions. The persistent process of enfolding and unfolding creates a particlelike structure analogous to a ball moving on a trajectory through space. To carry the metaphor a bit further, suppose we designate a small region of the body of water as point X and visualize a cork floating at this point. When the wave packet comes through, the motion of the cork will be perpendicular to the wave front. As the packet comes into point X, the information it carries will be enfolded there; and as the packet leaves point X, its information is unfolded from there. With this metaphor, motion is seen as a series of enfolding and unfolding waves. (This concept will be discussed more fully under The Holographic Metaphor and in Chapter 4.) That is, from the point at which a particle (like the cork) is observed, a wave spreads out from its source, and another one converges into the point. In essence, one wave gives rise to the other.

The Copenhagen version of reality was based on the correlation of observations that manifested the particle (matter) from its background. What happened in between the observation and the manifestation was unknown. Wheeler remarked, "What we call reality consists of a few iron posts of observation between which we fill in by an elaborate papier-mache construction of imagination and theory."[20]

Bohm, in his hidden variable paper, proposed a new model of motion that had profound implications for conceptualizing reality. He modified the Hamilton-Jacobi approach so that it conformed to quantum mechanics and introduced a new term called the "quantum potential" (discussed in the following section). He reformed the Schrödinger equation by dividing it into two parts: the first described the classical particle, and the second provided for the quantum potential. The quantum potential is not a classical potential such as electromagnetic or gravitational potentials. The information coming in and out of point X as the wave packet passes through is identified with the quantum potential and is determined by the enfolding and unfolding process. The quantum potential allows for the nonlocality requirement of Bell's theorem and leads directly to Bohm's concept of wholeness. Since Bohm's interpretation of the Schrödinger equation is just a mathematical transformation, its application is identical to standard quantum theory. Its philosophical interpretation is far different, however, and this novel approach suggested by Bohm in the early 1950s was largely passed over by the physics community.

Bohm and Krishnamurti

In 1959, Bohm came across a book called *First and Last Freedom* by the Indian philosopher Jiddu Krishnamurti. In it Krishnamurti discussed the concept of the observer and the observed, stating that a distinction could not be made between them. Bohm recognized the parallel to quantum theory. While Krishnamurti was referring to the psyche, Bohm sensed a connection to contemporary physics.[21] Two years later, Krishnamurti visited London, and he and Bohm had extensive discussions. The long and fruitful relationship that developed between them was part of what stimulated Bohm to expand his thinking beyond the realm of physics.

Krishnamurti typically opposed traditional teachings and authorities. He spent sixty years traveling about the world, lecturing and holding discussions with scholars, artists, and scientists, including Bohm, with whom he jointly authored several books. In *The Ending of Time*, a collection of dialogues between them in the early 1980s, complicated concepts unfold impressively, developing in a natural way from the two men's questions and answers to each other. These conversations seemed to nurture the growth and eventual flowering of their fundamental thoughts on the nature of reality.

After writing many articles and books in and out of physics proper, Bohm published *Wholeness and the Implicate Order*, a collection of essays summing up his thinking until 1980. This book presents a new conception of order in which the nature of reality and consciousness are described as a coherent whole, consisting of an unending process of enfoldment and unfoldment from a hidden level called the implicate order. The implicate order, the underlying order from which the wave unfolds, is the nonmanifest aspect of reality; it is outside space-time. Because of his understanding of quantum theory and his conversations with Krishnamurti, Bohm rejected the notion that, as he phrased it, "one who thinks is, at least in principle, completely separate from and independent of the reality he thinks about."[22] His hidden variable theory evolved into his concept of the implicate order being engaged in a process of enfoldment and unfoldment. This concept, in turn, led to the postulation of a meaning or consciousness throughout nature, which Bohm called "soma-significance" (discussed later in this chapter).

Bohm and Mathematics

Today, much of theoretical physics consists of forays into esoteric mathematics, often without sufficient physical underpinnings. Most theoretical physicists who are acquainted with Bohm's model either reject it or are uninterested because it lacks the explicit mathematical relationships to capture the attention of the physics community. It is true that Bohm's discussion of the implicate order is more qualitative than quantitative. However, Bohm's concepts offer a solid and deep philosophical foundation to modern physics, and until his death Bohm and his co-workers continued to work toward mathematical formulations of these ideas.

When Bohm was a student, most physicists felt that the intuitive concept was most important and that the mathematics would be developed in relation to it. But as quantum theory came to exercise more influence on physicists' thought, this attitude began to shift. Bohm sums up this change as follows:

> It was really because the quantum theory, and to a lesser extent relativity, were never understood adequately in terms of physical concepts that physics gradually slipped into the practice of talking mostly about the equations. Of course, this was because equations were the one part of the theory that everyone felt they could really understand. But this inevitably developed into the notion that the equations themselves are the essential content of physics. . . .
>
> Now I don't agree with these developments. In fact I feel that the current emphasis on mathematics has gone too far.[23]

While Bohm concedes that mathematics can give rise to creative insights, it is rarely the sole contributor to scientific discoveries. He cites Einstein's work as an example. Although Einstein used and appreciated the mathematics involved, he did not begin with it. During his most creative period, he started with a succession of images, which he later developed into more detailed concepts. Einstein had this to say about the role of mathematics in physical theory:

> Fundamental ideas play the most essential role in forming a physical theory. Books on physics are full of complicated mathematical formulae. But thought and ideas, not formulae, are the beginning

of every physical theory. The ideas must later take the mathematical form of a quantitative theory, to make possible the comparison with experiment.[24]

In the words of Bertrand Russell, "Physics is mathematical not because we know so much about the physical world, but because we know so little: it is only its mathematical properties that we can discover."[25]

THE CAUSAL INTERPRETATION: AN EARLY VIEW

Bohm's hidden variable theory was called the causal interpretation of quantum mechanics and represented the first step toward his more revolutionary concept of the implicate order. These early views had a deterministic flavor, hence the name "causal." But from the causal interpretation arose the idea of nonlocality, a concept with little appeal to the physics community since it was definitely a move away from the classical perspective. Later Bohm applied his causal approach to the quantum field, which led to the ideas that we will discuss.

Bohm's causal intrepretation can be seen as a proposal that the particle is *both* a wave and a particle, present together, and engaged in some sort of interaction. He began with the idea of a real particle surrounded by a quantum field that obeys Schrödinger's wave equation. The particle and field are never separate. Since the wave field is as important as the particle aspect to the description of, say, an electron, both are needed to fully understand it.

The Copenhagen interpretation regards the particle as having both wave and particle characteristics, which, being mutually exclusive, give rise to the idea that an observer must be present to cause one form or the other to manifest. In Bohm's view, however, the wave function is no longer merely a probability wave — a mathematical symbol, as in the Copenhagen view — but has a reality of its own, albeit outside space-time. Perhaps the most appropriate way to describe the elementary particle is to say that the wave and particle are *two* aspects of *one* new kind of entity, which is neither wave nor particle. Bohm has suggested that the electron is an entity that continually fluctuates from a particlelike character to a wavelike character and then back again. This is explained in more detail in Chapter 5.

Normally, in classical mechanics, the particle moves according to

Newtonian laws of motion while under the influence of forces derived from classical potentials (electromagnetic and gravitational force). The new Schrödinger field creates an additional potential called the quantum potential, corresponding to the quantum field, which represents the state of the whole. The quantum potential acts on the entire quantum system and is *not* directly determined by any single section of the system. The quantum potential thus relates to the particle *and* the field, or to the two aspects of the wave-particle duality.

To form a visual image of the quantum field, we need to understand that the classical physicist saw the world as made up of two elements: matter (particles) and fields (waves). The two types of fields, electromagnetic and gravitational, created a distribution of forces throughout space, which acted upon matter in definite ways. Newton outlined the action of the gravitational field, and Maxwell described the electromagnetic field. Quantum theory brought about a union of fields and particles. For lack of a better name, this combination has been called "quantumstuff."[26] Since the quantum field is not directly observed, it has been used mostly as a calculating device, and its reality has been open to question.

As noted earlier, Bohm saw the similarity between the Schrödinger equation and the Hamilton-Jacobi equations. By making a certain mathematical approximation, Schrödinger's equation becomes equivalent to the Hamilton-Jacobi equations. But if this approximation is not made, an extra potential results that represents a new type of "force" acting on the particle. Bohm labeled this "quantum potential," indicated by the symbol Q. The quantum potential is extremely sensitive and complex, producing what is seen as quantum chaos. Prediction of the movement of a particle acted upon by Q seems to be all but impossible.

Classical Potentials and the Quantum Potential

The idea of potential is extremely useful in physics, where it is applied mainly as a calculating device. To conceptualize potential, however, to know what it "really" is, has proved extremely difficult. For example, even ordinary electricity eludes our intellectual grasp. This elusiveness is amusingly characterized in the following exchange between an Oxford teacher and student:

Examiner: What is Electricity?

Candidate: Oh, Sir, I'm sure I have learnt what it is — I'm sure I *did* know — but I've forgotten.

Examiner: How very unfortunate. Only two persons have ever known what electricity is, the Author of Nature and yourself. Now one of them has forgotten.[27]

The term potential originated with Newtonian physics. The idea that things moving through space influenced each other at a distance, not just through collisions, was basic to the classical view. Out of that notion of action at a distance came the concepts of force and potential. Used mathematically, potential gives very accurate predictions of phenomena. Attempts to create a visual model of potential were unsuccessful, but because the mathematics worked so well, physicists were not too concerned. (To carry the electromagnetic force or potential, a pervasive ether was postulated. The ether was encumbered with several unconvincing attributes, however, and Einstein proposed in its place the deformation of space-time, which means, again, that the mathematical results were breathtaking, but the visualization was another matter.)

Classical potentials are defined as conditions of space capable of causing physical events. The potential energy is related to the work an object can do because of its position or state. Potentials are used in physics as intermediaries between objects that are spatially separated. Classical potentials have no existence without a source. The source (normally through movement) propels itself into the surrounding space and thereby affects other objects.

Several attributes differentiate the quantum potential from classical potentials. First, the quantum potential does not have a source within the three-dimensional world. It arises neither from the movement of a charged particle nor from an object that has stored gravitational potential, nor does the field from which it is derived radiate like an electromagnetic field. Second, the quantum potential does not have motive power. It does not force the particle to move, but is similar to an information field in that it carries information telling the particle *how* to move. Third, the relationship between any given particles goes beyond just these particles alone. The quantum potential is dependent on all the particles in the system — in fact, on all the particles in the universe. And for Bohm, the quantum potential leads to a new notion of unbroken wholeness. Any separation of systems into independent elements can be described only in an approximate way.

> We have reversed the usual classical notion that the independent "elementary parts" of the world are the fundamental reality and that the various systems are merely particular contingent forms and arrangements of these parts. Rather, we say that inseparable quantum interconnectedness of the whole universe is the fundamental reality and that relatively independently behaving parts are merely particular and contingent forms within this whole.[28]

This view calls into question the validity of a space-time continuum as being the foundation of reality.

Another attribute of the quantum potential is that it is multidimensional. Ordinarily a many-dimensional field is very difficult to visualize, but since the quantum potential arises out of an informational field, in which information is organized in many dimensions or multidimensional patterns, the concept of multidimensionality may be a little more accessible (see Chapter 7 for a discussion of Hilbert space). Mathematician Rudy Rucker comments on space and dimensions:

> It can be argued that the whole question of the dimensionality of space is a bogus one. What we really have is an influx of information, and we arrange this information in certain ways. Certain kinds of arrangements are best thought of in spatial terms, but all that's really there is a lot of bits and a mnemonic system for remembering which bit means what.[29]

An increase in dimensions means an increase of information, resulting in a more precise formulation. The construct of geometric multidimensions is certainly not unheard of in physics. Physicists have used such models as mathematical devices without hesitation. But they are not considered "real."

In this century, there has been much conjecture about "real" extra dimensions. One of the earliest speculations was by the Polish physicist Theodor Kaluza. He formulated a five-dimensional universe to unite electromagnetic forces and gravitational forces, but he was unable to explain quantum phenomena. Later a Swedish physicist, Oskar Klein, provided an explanation of why we do not "see" the effects of a fifth dimension in our macroscopic world. His explanation is rather simple: the fifth dimension is rolled up into an exceedingly small measurement of about 10^{-33} cm.[30]

At present, there is no experimental apparatus to confirm these small dimensions. Literally anything goes at this level, because quantum fluctuations start affecting space, and all physical laws are suspended. Nevertheless, the concept of a multidimensional field is fairly common and is no longer considered bizarre.

Bohm states that our three-dimensional reality contains aspects of a ground that is based on more dimensions than we normally perceive — that our everyday world is a three-dimensional projection of a higher-dimensional reality. He uses the "fish tank" analogy to help us conceptualize multidimensional space. Imagine a fish swimming about in a rectangular tank, with video cameras placed on two perpendicular sides of the tank. The filmed images are projected side by side on two video monitors, where we seem to see two fish swimming in a correlated fashion. Actually, there is only one fish. Projecting the three-dimensional reality upon two-dimensions creates the illusion of two fish. Similarly, the idea of a quantum potential (multidimensional information potential) suggests that what we see as separate parts of reality are only aspects of a totally interconnected underlying quantum world. This leads us to the notion of a seamless whole as the fundamental reality.[31]

The important point to note is that in his early causal interpretation Bohm considered the wave that determined the quantum potential as real; that is, it had its own actuality and was not just an algorithm. When the quantum wave was interpreted as a probability wave, its intensity was proportional to the probability that a particle was actually there. In the Copenhagen interpretation, the particle comes into existence only when a measuring device is applied. Bohm felt that the probability indicated the chances of the particle being there with or without a measuring instrument. This is a subtle but important difference.

According to Bohm, the wave function took on a double meaning. The first was a function from which probabilities could be determined, and the second was a function from which the quantum potential could be determined. Using this approach he was able to duplicate the results of quantum theory. The relationships were completely causal, and there was no need for the collapse of the wave function. Bohm's causal interpretation created a quantum potential that was a function of all the particles, and its intensity did not fall off with distance. Those features indicated a nonlocal connection, which, superficially at least, collided with relativity theory. The contradiction disappeared when Bohm applied his causal interpretation to the quantum field. In this way he developed a series of orders (implicate), a concept that will be discussed later in this chapter.

Bell found Bohm's causal interpretation very interesting. He saw that Bohm's "fully worked out hidden-variable account of quantum mechanics in which everything was deterministic and definite" proved incorrect mathematician John von Neumann's contention that hidden variables are not possible:

> Bohm's paper wasn't rigorous. It didn't have big displays of axioms, theorems, or lemmas [corollaries]. But one could see immediately that what he was saying was right. My reservation about his work and that of others in the physics community was that it was nonlocal, that what you do here has immediate consequences in remote places. And that was extremely odd.[32]

The Quantum Potential as Information Field

One of the significant features of Bohm's quantum potential is that its effect on the particle depends on its form rather than its magnitude. This irrevocably separated the quantum potential from any classical potential and led to a new concept of the nature of matter. In the classical world, a wave's encounter with an object creates a force on the object that is normally proportional to the magnitude of the wave. With the quantum potential, the effect is the same regardless of the strength of the wave. The wave may have large effects even at long distances, for the wave does not carry energy; it is an information wave.

Bohm uses a metaphor to illustrate his concept of the quantum potential and the changes it requires in our understanding of matter.

He suggests that we imagine a ship moving along with an automatic pilot. The ship sails on its own power but is guided by radio waves. The effects of the radio waves are independent of their strength but are dependent on their form. The information contained within the radio waves actually guides the enormous energy possessed by the ship. Information is fed into the shipboard computer, which then adjusts the direction and speed of the ship according to the data being processed.

When applied to the particle, this analogy indicates that the particle moves via its own energy but is directed by the quantum potential. If the quantum field is in a rapid and random state of motion, that information is conveyed to the particle. If all classical forces are removed from the particle, the particle still seems to exhibit random movement. The quantum fluctuations can be regarded as coming from a deeper, subquantum level; the motion of the particle is controlled by the underlying quantum potential. The information in the quantum potential is very detailed, causing the trajectory of the particle to appear chaotic. Actually, the apparent chaos is the result of the complexity of the quantum potential.

In the case of the ship, the power necessary for movement comes from its own engines. Where does the motive power for the particle come from? That question was posed to Basil Hiley, a collaborator of Bohm's. His response was:

> We always have the zero point energy. We know the vacuum state is actually full of energy, and the orthodox theory exploits that energy.

> But I'm not thinking of this in terms of the electromagnetic background, because the quantum potential arises from a field that is not like an electromagnetic field. It seems to be very different; it seems to be much subtler than that.[33]

Zero-point energy is an irreducible quantity of energy which, according to quantum mechanics, always resides in a system that is confined. The existence of this residual energy came about as a consequence of Heisenberg's uncertainty principle.[34] The zero-point energy that Hiley refers to is associated with the quantum potential field rather than with a classical field such as the electromagnetic field.

The analogy of the ship with its radio waves highlights an important concept in Bohm's causal approach. If the electron is guided in the same way as a ship with computerized guidance capability, how does such a "simple thing" as an electron perform any function comparable to the ship's computer? After all, the particle is thought of as a glob of quantumstuff, basically structureless. How can such a particle respond to the information carried by the quantum potential? Bohm's response is that there exists an inner complexity to matter. We are accustomed to the notion that as we probe deeper into smaller and smaller pieces of matter, the behavior becomes simpler or more mechanical, but this may be incorrect. On the contrary, in Newtonian physics simple mechanical laws are based on large aggregates of particles. And in society, the behavior of individuals is much more diverse and elusive than that of groups (which obey statistical laws). Thus, the behavior of the quantum world is likely to be far more subtle and complex than we realize.

It is true that experiments have failed to uncover any structure at the level of 10^{-16} cm, the approximate limit of current instruments. Between 10^{-16} and 10^{-33} cm there are still a lot of possibilities where structure might be discovered, although that range — equivalent to the size difference between ourselves and that of an elementary particle — is a no-man's-land for physicists. The problem lies in the technology used by experimentalists to "see" into these regions. Their "microscopes" are particle accelerators, and the higher their energy, the smaller the structures they can study. One of the implications of the uncertainty principle is that the energy (or resolving power) of the accelerator is inversely proportional to the area of structure to be studied. To separate or disintegrate a given mass into its constituent parts requires ever-increasing energies. To see much smaller than 10^{-16} cm would require accelerators of enormous expense. To see down to 10^{-33} cm would require technology not now available at any cost. The problem of the lack of technology is compounded by the distressing fact that the subquantum world cannot be treated even theoretically. At these infinitesimal dimensions, gravity again becomes the dominant force (in particle physics, gravity is not strong enough to demand consideration), and no quantum theory of gravity has been developed.

In any case, this region, according to Bohm, would need to reveal enough structure to respond to information.

> As you probe more deeply into matter, it appears to have more and more subtle properties. . . . In my view, the implications of physics seem to be that nature is so subtle that it could be almost alive or intelligent.[35]

Physical chemist Ilya Prigogine, in *From Being To Becoming*, also comments on the complexity of elementary particles:

> When I speak of a scientific revolution, I have in mind something different, something perhaps more subtle. Since the beginning of Western science, we have believed in the "simplicity" of the microscopic — molecules, atoms, elementary particles. Irreversibility and evolution appear, then, as illusions related to the complexity of collective behavior of intrinsically simple objects. This conception — historically one of the driving forces of Western science — can hardly be maintained today. The elementary particles that we know are complex objects that can be produced and can decay. If there is simplicity somewhere in physics and chemistry, it is not in the microscopic models.[36]

What does Prigogine mean by the statement that elementary particles are complex objects? He states:

> I am . . . convinced that we are only at the beginning of a deeper understanding of the nature around us, and this seems to me of outstanding importance for the embedding of *life in matter as well as of man in life.*[37]

Perhaps as they pursue deeper regions of space-time and matter, scientists will discover that what now appears as ambiguity is actually an enormous complexity with its own internal order. This notion will be more fully explored as we unfold Bohm's notion of order. Also, in Chapter 5 we will discuss phenomena occurring in the 10^{-33} region that may open the door to the implicate order. Physicist Chris Isham notes that this region may reveal surprises.

> For many workers, the strongest motivation for studying quantum gravity has always been this expectation that something "odd" happens at the Planck length [10^{-33} cm], coupled with the belief that understanding this "something" will involve a fundamental reappraisal of our basic concepts of the physical world, including perhaps the downfall of both general relativity and quantum theory.[38]

The idea of a quantum potential affecting a particle through its form leads us to the evolving concept of information. The *Random House Dictionary* defines information as "knowledge communicated or received concerning a particular fact or circumstance; news." In the Middle Ages the word had a different meaning: information was not just a report of an event but also accounted for the results caused by the report. It was more like a force that guides events; it was more active. Technological developments during World War II resulted in a redefinition of information that is closer to the medieval concept of information as an active agent. Jeremy Campbell, science writer and correspondent for the *London Evening Standard*, sums this up in *Grammatical Man*, discussing the evolution of information theory from World War II to the early 1980s.

> Information emerged as a universal principle at work in the world, giving shape to the shapeless, specifying the peculiar character of living forms and even helping to determine, by means of special codes, the patterns of human thought. In this way, information spans the disparate fields of space-age computers and classical physics, molecular biology and human communication, the evolution of language and the evolution of man.

> Evidently nature can no longer be seen as matter and energy alone. Nor can all her secrets be unlocked with the keys of chemistry and physics, brilliantly successful as these two branches of science have been in our century. A third component is needed for any explanation of the world that claims to be complete. To the powerful theories of chemistry and physics must be added a late arrival; a theory of information. Nature must be interpreted as matter, energy, and information.[39]

Note that Campbell sees information as giving form or shape. It is the source for order in the universe; it is the formative cause. Along with matter/energy (which relativity theory showed to be equivalent), information has become increasingly recognized as a major constituent of nature, no longer seen as merely controlling a process. This meaning is also closer to the way Bohm envisions the quantum potential as a conveyor of information. Bohm implies that an electron must have a "structure" to receive this information, and he looks for it in the realm between 10^{-16} cm and 10^{-33} cm.

Later in this chapter and in subsequent chapters, we consider the idea that the universe is not made of matter/energy and information, but is composed of information only, or some variant of it.

THE HOLOGRAPHIC METAPHOR

The idea of information carried by the quantum potential leads to Bohm's fundamental concept of wholeness. The quantum wave's information is potentially everywhere, since the form is not dependent on distance. Bohm calls the information used by the particle active information. At other points, it is potentially active information. The information includes the state of all particles within the whole. When a particle taps into the information field, its movement reflects the state of the whole. In that sense, the particle cannot be separated from its environment and cannot be treated as an independent entity; it is an aspect of the entire system. Classically, the whole is the sum of its parts. With the quantum potential, the whole *organizes* the parts. To restate this, the essential feature of the quantum field and the resulting quantum potential is that any event happening anywhere is immediately available everywhere as information. Each portion of space contains information about all portions of space. The reason the classical world appears to us as separate and distinct objects is because the quantum potential at the macro level is negligible (with some exceptions, such as superconductivity). We are left with a three-dimensional projection from a multidimensional ground.

As Bohm sees it, all information is enfolded within each region and is therefore available to all particles, and all quantum fields of all particles have an input in every area. Bohm used the hologram as an analogy for space where these quantum fields combine. (A note of caution: the hologram metaphor is static, classical, not multidimensional, and therefore limited. A more dynamic conception followed with Bohm's introduction of quantum field theory and its extension into the superimplicate order.)

A hologram is a photographic record of light waves reflected by an object. The object is illuminated with laser light. (Normally laser light is used, but other radiations such as sound can also produce holograms.) The reflected light from the object is then combined with the direct light from the source. That combination produces an interference pattern on the photographic plate. A three-dimensional image of the object can then be produced by illuminating the interference pattern with the original laser light.

The pattern carries all the visual information related to the object. That information is contained in the phase, direction, and intensity of all the light falling on the photographic plate. The significant aspect of the hologram for our discussion is that each region of the photographic plate carries information about the whole object. If a small region of the plate is illuminated, the three-dimensional object still appears, though it is somewhat fuzzy. In holographic photography, there is not a one-to-one correspondence between the points on the plate and the points of an object as in normal photography. Each region in holography is relevant to the whole structure, because light from all points on the object is enfolded into each area of the plate.

To Bohm, the hologram provides an analogy for movement that is consistent with quantum mechanics. To help us better understand this concept, Bohm describes the process we go through when we look at something. Light from each point becomes enfolded in the eye, bringing a tremendous amount of information into the small area of the pupil. This information is then unfolded through the lens of the eye, sensory nerves, and the brain, literally creating an order that appears three-dimensional.

Bohm mentions a second analogy that shows how enfoldment and unfoldment are essential to scientific observation. When we use a telescope to observe the universe, the light coming from all the stars is enfolded into the small region of the telescope lens. One might expect these waves to produce a totally chaotic pattern. Yet the small area of the telescope lens is able to enfold the whole visible universe in such a way that it unfolds meaningfully to the observer. This is similar to a hologram in which light from all points of an object is enfolded in each small area of the holographic plate.

With these analogies as background, Bohm offers a picture of enfoldment and unfoldment at a particular point in space. When a wave front approaches point X, the information contained on the surface of the wave front is enfolded into this point, or rather in the wave function at this point. The wave function in turn determines the quantum potential, which supplies information to the elementary particle at point X. Therefore, all wave fronts coming into point X pass information on to the particle at that point by the process of becoming enfolded into X. Also, there is an unfolding from point X in the form of a wave that also carries information. The quantum potential that informs the particle is then the result of the enfolding and unfolding processes at point X. Like a hologram, point X contains information from all other points, and all other points contain information from point X.

This example of unfolding and enfolding can be extended to a particle in motion. If we describe the motion of a particle from point A to point B, in the Copenhagen approach, the particle at A can also be seen as a quantum wave spread throughout space. By calculating the wave function for this experimental situation, the probability for the particle ending up at B can be determined. If a measuring device is placed at B and a particle is manifested, the quantum wave collapses from its spread-out position to a spike at point B. That probability is now actualized, and all other probabilities have gone to zero.

The process is viewed quite differently by Bohm as arising naturally from the quantum mechanical formalisms. A wave leaves point A, or is unfolded. It then is enfolded at a second point. The second point unfolds a wave, which is enfolded at yet another point. All points are reached by waves from all other points. By this process, all the waves starting from A reach B through a large number of intermediate steps. The total wave at B represents all the waves that have come from A through all possible paths from A to B.

Bohm's metaphor is based on a unique approach to motion developed by physicist Richard Feynman in 1950. Feynman proposed that a particle not observed is a wave, and as such it takes *all* possible paths from A to B. In this sum-over-histories approach (as Feynman called it), the wave arriving at B is the sum of all contributions of all possible paths from A to B. The resultant path is the one taken after all interference effects have occurred. Even though Feynman talked about "paths" of elementary particles, the diagrams dealt only with waves that produced interference effects. So the particle really could not be seen as following a path. Although Feynman's approach did not visualize both the particulate and wave aspects of the elementary particle, it nevertheless became an invaluable method for handling difficult calculations. Bohm saw in the Feynman diagrams an imaginative picture of motion, once the concept of unfoldment and enfoldment is added.

To visualize this process as the path of a particle, we must remember that an electron has both a particle and a wave aspect. The trajectory of a particle can be viewed as a series of points, each displaying the basic movement of enfolding and unfolding. The quantum potential at each point determines the process and in turn guides the particle along its apparent path from A to B. All the information in the environment is available to the particle through the quantum potential.

This approach definitely does not conform to the description of motion as little billiard balls moving through space. Rather, the particles here are dynamic entities, grounded in the whole and projected and injected from this ground. A continual process of unfolding and enfolding produces an object that displays the properties of a particle moving on a trajectory through space. In Bohm's view, particles are not separate entities that interact; their relationship is not something outside of the particles themselves. He sees all mutual effects as mediated by an underlying whole.

When motion is viewed in this way, the space-time continuum is no longer considered basic. Physicist Jack Sarfatti offers as an analogy for quantum motion a strip of lights timed to turn on and then off, one after another, creating the illusion of movement along the strip. This example shows how a particle *appears* to move across space-time — how the universe displays the illusion of continuity. If the "explications," as Bohm calls them, unfold with greater separations between events, then the particle seems to jump from one place to the other, displaying the discontinuous nature of the universe. An example of a discontinuous series of states is an electron jumping from one orbit to another in an atom.

ORDER: AN INFINITE SPECTRUM

Defining Order

In *Science, Order, and Creativity* Bohm attempted to clarify his concept of the implicate order by developing a general view of order.

This in turn led him to the idea of the "generative order," which includes the implicate order as a special case. But first let us look at Bohm's idea of order in its broadest perspective.

The first notion of order we have is through our ability to recognize similarities and differences, a facility residing in our senses, especially vision. Our perception proceeds first by gathering differences; these data are then used to construct similarities. Once this is done, categories can be formed. Certain things are selected by our mental process as being different from some general background. These things can be brought together, and we can regard the differences between them as unimportant while viewing their common difference from the background as significant.

As an example, if we were walking in a park that contained several gardens, we could extract the gardens from the background and place them into a common "garden" category, ignoring the differences between them. They differ from the park in that they contain a variety of smaller plants in distinct arrangements. The category "garden" applies to all of them and provides enough information to allow us to extract them visually from the background.

Categories are not final and static; they are dynamic. That is, as some differences assume greater importance and new similarities are collected, new categories are formed. For instance, the gardens themselves exhibit differences: one is a flower garden, another a vegetable garden. With the formation of these new categories, we have created two levels of classification. On one level we have gardens in general, and on a second level we have flower gardens and vegetable gardens. The new classification occurs because of a change in context. By knowing and recognizing the differences between flowers and vegetables, we are capable of creating new categories, or orders.

As Bohm points out, classification is linked inseparably to the idea of a dynamic order. Furthermore, the classification, or creation of order, takes place both through the senses and the mind. The senses pick up the differences between a flower and a vegetable, and the mind, through a creative action, categorizes the differences. All this implies a spectrum, or levels, of order, and that spectrum is clearly dependent on context.

Bohm offers analogies that help clarify the concept of a spectrum.

The spectrum can be thought of as a series or continuum of degrees of order. As an example, if a ball is rolled down a hill that is perfectly smooth and inclined to the horizontal at a definite angle, it will go down in a straight line. If the initial velocity and the initial position of the ball are known, then it is possible to correctly define its trajectory. Bohm calls such a situation an order of the second degree, because two pieces of information are required to establish the level of order.

If the incline is changed so that it contains a series of bumps, the motion becomes complex, and much more information is needed to define the trajectory. Thus, the degree of order increases. If each irregularity is carefully defined and described, in terms of physical laws, then the order can be reduced to the second degree. Again, since we only need the initial velocity and position to define the trajectory, it becomes a second-degree order. As previously stated, the difference between the higher order and the lower order is context-dependent.

A second analogy more clearly indicates that randomness and orderliness can be viewed as different points on the same spectrum. If a computer generates random numbers, an observer sees the readout as a series of numbers succeeding each other in a very complex and unpredictable way. By any standard, these numbers would be viewed as random, and the order certainly would be toward the random end of the spectrum. However, suppose the computer program that generated the numbers is added to the readout. Then the apparently random set of numbers becomes an order of a lower degree merely by adding new information to the context. In this way, the context has been changed. Likewise, a coded message will seem completely disordered until one knows the code, whereupon the message suddenly becomes ordered and meaningful.

Bohm considers the spectrum of orders to be infinite. He feels that there is no condition in which all information is present or there is a final context. In other words, there are no closed systems. No matter how far out the spectrum we go, randomness gives way to orderliness as broader contexts are included. Proceeding in the other direction along the spectrum, we again encounter randomness followed by orderliness if the context is expanded. In a sense, determinism and chance exist side by side, related by context.

Bohm sums this up:

It is clear then that *wholeness of form* in our description is not compatible with *completeness of content*. This is not only because subsystems may eventually have to be regarded as constituted of subsubsystems, etc. It is also because even supersystems will ultimately have to be seen as inseparable from supersupersystems, etc. This form of description cannot be closed on the large scale, any more than on the small scale. Thus, if we supposed that there was an ultimate and well-defined supersystem (e.g., the entire universe), then this would leave out the observer and it would break the wholeness, by implying that the observer and the universe were two systems, separately and independently existent. So, as pointed out earlier, we do not close the description on either side, but rather, we regard the supersystems and subsystems as ultimately merging into the unknown totality of the universe.[40]

Bohm does not accept the existence of a basically random state, or an end to the spectrum. Niels Bohr, for example, believed in an inherent ambiguity in the quantum world. But Bohm's view precludes irreducible randomness. In each succeeding broader context, meaningful order is always possible. On the other side, there is no complete determinism, because there is no end to the spectrum. There will always be hidden orders, and no "final" theory will ever be possible. We can only hope to keep extending our theories as we travel down the infinite spectrum of orders.

Perhaps Bohm's notion of order is best summarized by viewing the spectrum as information. Randomness can then be defined as a system in which the available information is incomplete. When the missing information is supplied by broadening the context, then randomness is converted to a given degree of order. The original information, which was incomplete, becomes a special case of the broadened system. The information of a random system is valid within the more restricted context.

A determinist would say that if all the information for the entire universe were available, then the laws of physics would be perfectly accurate. That would be closing the spectrum on one end. Bohr closes the spectrum on the other end by saying that there is a final ambiguity (randomness) that can never be made orderly by more information — in effect, a kind of quantum randomness. The essence of Bohm's view is that the information spectrum is infinite on both ends of the continuum. A physical system never contains *all* the information necessary for a complete and detailed description.

An analogy is that we are climbing a mountain of information, a mountain infinitely high. On each plateau, a broader vista is available (i.e., there is more information), but we can never see everything (information is never complete). Of course, we must be careful with analogies. Spatial separation is not implied in Bohm's hierarchy of orders; ascending the mountain of orders is rather a matter of broadening our focus. We might say that the mystic is on a higher plateau, that the mystic's view is broader but not as detailed as the physicist's from a lower plateau. But the mystic's information is also incomplete, because that level of order is also a subset of a greater truth.

Before we leave this general discussion of order, let us examine the relationship between a spectrum of order and the dimensionality of space. As mentioned earlier, Rucker sees in the dimensions of space a method of organizing information. That is, information can sometimes be arranged in spatial terms. If we try to describe a system and we have ten pieces of information, we can define it in a ten-dimensional space, indicating a correlation between the number of dimensions and the amount of information we possess. A view of an event in three-dimensional space is comparatively restricted. By increasing the dimensions, the view is more encompassing. So when mystics penetrate multidimensional realities, in essence they enlarge the context.

In summary, there appears to be an infinite spectrum going from orders of a low degree to those of a higher degree. This nesting of suborders into larger orders contains and conditions the suborders. The suborders are not independent but are affected by the larger order and in turn affect the larger order. Whatever degree we are discussing, we cannot say it is random, for its meaning is always dependent on context.

Generative and Implicate Orders

Now we can examine Bohm's concept of a generative order. A generative order differs from the orders described above in that it does not represent processes already in nature or in the mind. The generative order uses similarities and differences to actually "generate" shapes and processes. The generative order can be seen in the mathematical theory of fractals, in sculpture, in painting, in music, and in other creative endeavors.

In all these generative orders a process of creation begins with some overall perception, which is unfolded into particular forms. As an example, the theory of fractals allows for the formation of shapes of extreme complexity. These shapes are created by choosing different simple base figures which, acting as generators, are applied on smaller and smaller scales. Configurations ranging from snowflakes to coastlines are generated in a straightforward fashion using this technique. The visual artist, as another example, works with an original idea and allows it to unfold into forms. In music, a similar process takes place.

Bohm defines the implicate order as a special case of the generative order. The implicate order contains degrees of enfoldment that can be unfolded in some limited way. That unfoldment generates what Bohm calls the explicate order.

In the past ten or fifteen years, a new study called the science of chaos has come into existence. One of the basic findings of this discipline is that chaos is enfolded in order, but even deeper, another chaos takes over with an even deeper order behind that. In the words of computer scientist Douglas Hofstadter, "It turns out that an eerie type of chaos can lurk just behind a facade of order — and yet, deep inside the chaos lurks an even eerier type of order."[41] And as Seth says:

> There is no such thing, basically, as random motion. There is no such thing as chaos. The universe, by whatever name and in whatever manifestation, attains its reality through ordered sequences of significances.[42]

Both these statements appear to confirm Bohm's premise regarding a series of orders.

The whole concept of subsystems and supersystems as part of an infinite yet inseparable totality clarifies one of the major paradoxes of our time. The two bulwark theories of the scientific community are quantum theory and the theory of evolution. The quantum theory informs us that our macro reality is based upon a near-infinite number of chance events. These constitute the apparently lawful universe in which we live. On the other hand, according to the theory of evolution, innumerable chance mutations are responsible for the extraordinary variety of species on our planet. At present, it is assumed that each chance mutation is absolutely independent, yet collectively they function in an orderly way. Each separate event defies law, yet as a group these same events create law. In short, order and reason seem to be based on randomness.

If we accept Bohm's view, however, there is no irreducible randomness. When we study the explicate order only, structure seems to arise from chance. But Bohm sees this as only an abstraction of a more encompassing order. So-called chance events actually reveal the pattern of the superorder in which the events are embedded. Each event is aware, or knows, or is informed, of the overall plan in which it participates.

To summarize, the implicate order, which is a special case of the generative order, can be seen as an order that exhibits degrees of enfoldment and contains the capacity to be unfolded in a limited way. The unfoldment (which is explicated or projected) becomes manifest as the explicate order.

THE HOLOMOVEMENT: CARRIER OF THE IMPLICATE ORDERS

The holomovement is described as the infinite spectrum of generative orders. Remember that the generative order includes the implicate order. The holomovement, then, is an infinite spectrum of implicate orders, an unbroken and undivided totality. In certain cases, we can extract or abstract particular aspects of the holomovement, but basically all forms of the holomovement, all aspects, merge and are inseparable. In its totality the holomovement is unlimited, undefinable, and immeasurable. It is the ground of everything.

The movement of the holomovement is not from place to place, but is more like a flux. In this case, the flux is an enfolding and unfolding, in contrast to the idea of an entity crossing space in time. (Recall the image of a strip of lights.) Within the holomovement, the connections exhibited by the whole have no relationship to local positions in the space-time continuum; they all exist within the concept of enfoldment. The implication is that there is not a direct causal connection between events, rather the relationship takes place in the holomovement. The hologram is a static recording of the holomovement, or simply an abstraction; the hologram analogy shows us how the holomovement might be displayed.

Bohm calls this display the explicate order. He described it in an interview with philosopher Renée Weber:

> *Bohm:* The holomovement is the ground of what is manifest. And what is manifest is, as it were, abstracted and floating in the holomovement. The holomovement's basic movement is folding and unfolding. Now I'm saying that all existence is basically a holomovement which manifests in relatively stable form.
>
> *Weber:* The flux arrested for the time being?
>
> *Bohm:* Well, at least coming to balance for the time being, coming into relevant closure, like the vortex which closes on itself though it's always moving.[43]

According to this view, when the implicate order unfolds, the explicate order displays. The explicate order is the ordinary world of experience. It is the unfolded portion of the holomovement, which displays to us an aspect of the implicate order. The implicate order, on the other hand, provides the commonality for matter, life, and consciousness. It is in the implicate order that matter and consciousness are basically identical, differing only in subtlety. (This is discussed in more detail in the section on soma-significance.)

In Bohm's view of an infinite spectrum of orders ranging from orders of low degree out to random orders, a random order at the far end of the spectrum becomes an order of a lower degree in a larger context. It seems reasonable to assume that the implicate order is followed by further orders along the spectrum — the superimplicate, the super-superimplicate, ad infinitum.

Bohm's Causal Interpretation Applied to the Quantum Field

Bohm saw the superimplicate order as an extension of his causal interpretation, applied to the quantum field rather than to the particle. In this way, the superquantum potential is related to the entire quantum field as the original quantum potential is related to the particle. The quantum field then becomes the basic reality, with the particle being a manifestation or aspect of the field.

In Bohm's view, quantum field theory suggests that empty space is a vast ocean of energy. Bohm establishes a relationship between that ocean of energy and the implicate order, which unfolds to form space, time, and matter. The theory of general relativity implies that space and time exist only because of the presence of matter, that matter creates the space it occupies. In other words, matter is a particular state of space. Without it, there would be no space and no time. The ocean of energy from which matter springs is not primarily in space and time at all, and therefore it is not recognized in the mathematical formulations of physics. As Wheeler notes, "If we're ever going to find an element of nature that explains space and time, we surely have to find something that is deeper than space and time — something that itself has no localization in space and time."[44] This "something" might be Bohm's implicate order.

This application of quantum field theory allows the physicist to describe the wave-particle duality in a more comprehensible way[45] because the field can also be seen as a discrete particle. This is not to be construed as a return to the classical concept of an atomistic reality on a quantum level. Rather, it is a confluence of fluctuations in quantum fields.

A field is a form of wave motion; that is, it varies from one point to the other and also changes with time. Any periodic wave motion can be described by a theorem developed by the mathematician Joseph Fourier. Essentially, Fourier's theorem states that a periodic wave, no matter how complex, can be seen as a unique sum of sine waves. Sine waves are the familiar waves we encounter on a lake, on a guitar string, or on an oscilloscope. They are a smooth oscillation which spreads throughout space. Sine waves are easy to use mathematically, and Fourier's theorem has become a powerful tool in physics. Since the field is a wave and can be described as a set of sine waves, it can also be seen as a set of harmonic (sinusoidal) oscillators. Actually, a free electromagnetic field is mathematically equivalent to an infinite number of harmonic oscillators. When the field is quantized, the energy states are discrete; that is, the energy states can be thought of as wavelike excitations spread out over a broad region of space. In another respect, however, the field is like a particle with a discrete quantum of energy proportional to the frequency. The particle then can be interpreted as energy residing in the field and can be removed or inserted in discrete amounts.

In the classical approach, without excitation, the energy of the field would be zero. Any excitation above this would be allowable. In quantum field theory, only discrete excitations are involved, and there is never a zero-energy state, as in classical mechanics. The non-zero state follows from Heisenberg's uncertainty principle. This principle does not allow an oscillating system to rest. If it were at rest, it would have exactly defined positions and momenta, and the principle says that is not possible, that the system must always tremble somewhat.

Space as a Plenum

The universe can be seen as a large number of interacting fields (according to Fourier's theorem), which manifest as particles interacting with each other. Quantum theory tells us that when the energy state is at the lowest possible level, the field has zero-point energy. Due to this minimum or zero-point energy, if we were to add up all the wave-particle modes of excitation in any region of space, the energy would be infinite, because an infinite number of wavelengths are present.

Through quantum theory and general relativity theory, we know that the gravitational field also exhibits these wave-particle modes with each having a zero-point energy. In the case of gravitation, instead of coming up with an infinite energy, Bohm proposes that there is a shortest wavelength that can be added since there is a certain length of space-time at which the measurement of space-time becomes totally undefinable. He says:

> You have to ask what would be the shortest length and there seems to be reason to suspect that the gravitational theory may provide us with some shortest length, for according to general relativity, the gravitational field also determines what is meant by "length" and metric. If you said the gravitational field was made up of waves which were quantized in this way, you would find that there was a certain length below which the gravitational field would become undefinable because of this zero point movement and you wouldn't be able to define length.[46]

This view is similar to one proposed by Heisenberg in the early days of quantum theory. In order to explain the infinities encountered in calculations, he proposed that space itself is quantized. Therefore, the electron is not a point. The space continuum is actually grainy, and there exists a universal minimum length.

Bohm estimates this undefinable length — the smallest possible piece of space — to be 10^{-33} cm. Using 10^{-33} cm as a minimum wavelength, the energy in one cubic centimeter of space is far beyond the total energy in the known universe. This fact suggests that empty space, usually considered a vacuum, is in fact a vast energy pool, and that matter is a small quantized, wavelike excitation of this pool, similar to a ripple on a large ocean. At present, physical theories avoid consideration of this immense energy background by only considering the *difference* between the energy of empty space and that of space with matter in it. It is this difference that is used in calculating the properties of matter.

This vast energy pool is ignored because present-day physical instruments cannot measure it. The only energy recognized by physics is that which is explicated from the pool, and the measuring apparatus is sensitive only to the features of the field that last a considerable length of time. If a feature fluctuates rapidly, it is not detected. The so-called empty space produces no effects since its fields cancel themselves out on the average. As a consequence, space appears empty to the physical instrument.

Bohm does not ignore the energy pool. He identifies it as an aspect of the holomovement and defines it in terms of the multidimensional implicate order. To Bohm space is no longer conceived of as a three-dimensional vacuum but a plenum. The entire universe of matter is seen as a comparatively small pattern of excitation. This pattern is relatively autonomous and gives rise to recurrent, stable, and separable projections into the explicate order, which is three-dimensional.

Fritjof Capra in the *Tao of Physics* also describes the particle as an explication of an underlying quantum field:

> The quantum field is seen as the fundamental physical entity: a continuous medium which is present everywhere in space. Particles are merely local condensations of the field; concentrations of energy which come and go, thereby losing their individual character and dissolving into the underlying field.[47]

Bohm cautions us to keep in mind that this sea of energy refers only to lengths greater than 10^{-33} cm. At lengths less than 10^{-33} cm there could be other domains. The notion of the implicate order is still limited by the critical length of 10^{-33} cm.

Wheeler has conducted theoretical studies of the structure of space at these very small dimensions, and his views are similar to Bohm's. He points out that the density of field fluctuation energy in empty space (vacuum) is approximately 10^{94} g/cm^3. That is 94 zeros to the left of the decimal point — not very empty! In comparison, nuclear density is approximately 10^{14} g/cm^3. A difference of 80 zeros to the left of the decimal point exists between the two densities. With this in mind, we can appreciate how negligible the density of the particle is compared to empty space.

Wheeler uses the metaphor of a cloud in the sky to illustrate this point: "A particle means as little to the physics of the vacuum as a cloud means to the physics of the sky." Wheeler, like Bohm, identifies the basic starting point for physics not as the elementary particle but as space itself. He comments:

> That vacuum, that zero order state of affairs, with its enormous densities of virtual photons and virtual positive-negative pairs and virtual wormholes, has to be described properly before one has a fundamental starting point . . .[48]

Bohm, of course, identifies space with the implicate order, an aspect of the holomovement. In this view all the matter in the universe can be seen as "clouds" floating in a vast ocean of energy, which is not in space-time. Einstein's ghost field and Schrödinger's empty wave begin to reappear in a different context.

The Superimplicate Order

By using his causal interpretation applied to field theory rather than to the particle, Bohm is able to incorporate other implicate orders into his explanation. The causal interpretation applied to the field leads to two implicate orders. The first is the quantum field itself; the second, the superimplicate order, is a superquantum wave function. The superquantum wave function is related to the original quantum wave function in the same manner as the original quantum wave function is related to the particle. We have, then, a superimplicate order, or superquantum wave function, that is similar to the first, implicate order, but far more subtle and complex.

The field equations of the implicate order are modified in such a way that they become nonlocal and nonlinear; that is, the quantum potential becomes nonlinear. It is this quality of nonlinearity that leads to quantum wholeness. All the particles in the system are coupled together; the entire system moves as a whole. The superimplicate order becomes the source for organization and creative activity in the first implicate order.

Bohm states that the implicate order should be seen as two levels of enfoldment. The first level is "empty space" or a high-energy pool with ripples on it called matter. The second level is a super-information field (superimplicate order), which organizes the first level into various structures. Bohm notes that the implicate order does not have the intrinsic capacity to unfold an order, being essentially passive. The mathematical description is that the implicate order is "linear." The superimplicate order produces nonlinearity in the implicate order and thus organizes it or unfolds it into relatively stable forms with complex structures. Bohm states that this is how the quantum mechanical field theory works:

> The first implicate order is like the field, and there is a super-implicate order which organizes the field into discrete units which are particle-like. Without that superimplicate order however, the field would just spread out without showing any particle-like qualities.[49]

The organizing source for the implicate order lies at a deeper or more subtle level. Bohm postulates an infinite number of levels — implicate orders, superimplicate orders, super-superimplicate orders, etc. — each organizing the level below, each in a more subtle dimension than the one below. All these implicate orders merge into an infinite-dimensional ground, or holomovement. Furthermore, Bohm sees each entity (electron, proton, etc.) as having its own implicate order. In fact, all sets of entities, all gestalts of matter, have their own implicate orders. For example, an arrangement of particles has an implicate order of its own, apart from that of each individual particle. Finally, *all* the implicate orders of *all* arrangements are aspects of the indescribable holomovement.

An excellent analogy for the process of explication and implication as detailed by Bohm is given by Alex Comfort in his book *Reality and Empathy*. Comfort asks us to imagine a computer game consisting of a television screen, a computer with controls, and a player. The screen is a display in two dimensions of objects that can be made to move, collide, explode, and so forth. The whole show appearing on the screen is analogous to our conventional, everyday three-dimensional world. The causal arrangements of what happens on the screen are prewired and implicit in the computer. As Comfort points out, the game could be played without the display, but that would require the player to make massive and cumbersome calculations. The screen allows the player to engage in the game more efficiently. With all the moves prewired within the computer, the controls are used to select events and display them in order to determine the course of action. Thus, viewing the happenings on the screen guides the player in making selections. We can see the parallels to Bohm's orders. The screen is like the explicate order; without the computer and the player it would be dark. The computer is like the implicate order; without the player, it would just sit in a passive state. The player initiates the action; like the superimplicate order, the player is the formative and creative force.

In his book *Synchronicity*, physicist David Peat points out that physical laws have always been treated as mathematical abstractions that describe but do not "cause" events. As such, they have no more effect on the operations of our three-dimensional world than the description of a sporting event has on the game. Peat quotes Wheeler's description of this notion:

> Imagine that we take the carpet up in this room, and lay down on the floor a big sheet of paper and rule it off in one-foot squares. Then I get down and write in one square my best set of equations for [how] the universe [works], and you get down and write yours, and we get the people we respect the most to write down their equations, till we have all the squares filled. . . . We wave our magic wand and give the command to those equations to put on wings and fly [i.e., make the universe work]. Not one of them will fly. Yet there is some magic in this universe of ours, so that with the birds and the flowers and the trees and the sky [the whole thing] flies. What compelling feature about the equations that are behind the universe [i.e., the actual laws of nature] is there that makes them put on wings and fly?[50]

Peat responds that "the equations of physics will never take wings and fly for they are simply mathematical descriptions." He goes on to suggest that mathematical laws are manifestations of something beyond our universe that is "creative, generative, and formative." This concept is similar to the organizing source for the implicate order, that is, the superimplicate order. The superimplicate order is what makes the laws of nature operate — what turns them on, so to speak.

In summary, Bohm changed his original mechanical approach, represented by the causal interpretation, by applying the causal approach to quantum field theory. He started with the classical idea of a continuous field, a field spread throughout space. He then applied quantum theory rules to the classical field and came up with quantized values for certain properties (energy, momentum, and angular momentum.) This information indicated that a field could act like a collection of particles or exhibit wavelike properties.

To reiterate, when Bohm applied the causal interpretation to quantum field theory, he found that the field could be modified by a superquantum potential, and this turned out to have the same relationship to the field as his original quantum potential had to the particle. The superquantum potential is the agent that brings together the field at one region to produce a manifestation that we call a particle. It also provides the stability needed to sustain the existence of the particle. So the original quantum field converges and diverges from one point to another, creating the appearance of particle motion, while the spreading wave retains its wavelike characteristics. The superquantum potential fashions form out of the quantum field, which gives rise to our manifest world.

Bohm identifies the quantum field with the implicate order. The superquantum potential he identifies with the superimplicate order, and he goes on to postulate a hierarchy of implicate orders. The second implicate order he calls a source of formative, organizing, and creative activity. The first implicate order is engaged in a process of rapid and constant creation and annihilation, which results in an explicate order containing relatively constant, recurrent, and stable aspects of the whole.

Characteristics of the Holomovement

It is important to emphasize three aspects of the holomovement and its implicate orders. These aspects, as we shall see, find common ground with the Perennial Philosophy and with Seth's description of reality.

1. The universe of matter floats in a vast ocean of energy. Contemporary physics starts its analysis with the particle and disregards the ocean. In effect, physicists regauge the energy spectrum so the pool is considered at zero energy. Bohm takes a different view of that energy pool in developing his concept of the holomovement.

2. This ocean of energy — the implicate order — is not inert. In some sense, it is conscious and alive. Since matter springs from the implicate order, it too is alive — not aware in the way of human consciousness, but having its own natural inclinations and tendencies.

3. The holomovement is a spectrum, a continuum of consciousness with matter on its lowest rung. (Bohm's views on consciousness will be considered in the section on soma-significance in this chapter.)

By using the concepts of both a hierarchy of implicate orders and the hologram in describing the holomovement, we are presented with a paradox. While consciousness is sometimes thought of as occupying various levels, the entire range of implicate orders is present at every level. According to the Perennial Philosophy, all levels of consciousness not only are described by stages of development, but also are considered as perfectly interpenetrated. Likewise, the holomovement is not to be visualized as a hierarchy spread out in space, but as having every point interpenetrated with every other point. The separation of objects is an attribute only of the explicate order, as are space and time. Therefore, the paradox of the hologram and the spectrum exists only in the explicate order.

The union of opposites is a basic tenet of Eastern thought; truth comes only through an amalgamation of two interacting opposites. As Bohr said, "The opposite of a fact is a falsehood, but the opposite of one profound truth may well be another profound truth."[51] Seth comes to a similar conclusion:

> There are profound complications beneath my words. . . . Opposites have validity only in your own system of reality. They are a part of your root assumptions, and so you must deal with them as such. They represent, however, deep unities that you do not understand. . . . The effect of opposites results, then, from a lack of perception. . . .[52]

Paradoxes as they appear in our three-dimensional world are in part the result of our attempt to describe reality with limited analogies. Many of our paradoxes are resolved when our view is broadened to include other orders.

TIME, A DERIVATIVE OF THE TIMELESS

Bohm's metaphors have important implications for our understanding of time. Before addressing his views, let us examine how present-day physics interprets time.

The physicist's conception of time is intimately related to the second law of thermodynamics, which states that as time goes forward, the universe becomes more disordered. This disorder is called entropy, and it is mathematically defined. The increase of disorder means that the universe is moving toward irreversible randomness and chaos.

In the latter part of the nineteenth century, physicist Ludwig Boltzmann explained the second law of thermodynamics on the basis of atomic theory. He saw entropy as the result of random collisions between atoms. This idea was placed on a sound mathematical footing, but there was a problem. The laws of motion of the individual atom follow Newton's laws of mechanics, in which time is reversible. That is, the laws do not indicate what is past and what is future; they are equally applicable in both directions.[53]

How does one go from a time-reversible situation on the atomic level to irreversibility on the macro level? The explanation was that the macro world is not "truly" irreversible. The second law, in effect, could be violated, though the probability of reversible time is so small as to be negligible for all practical purposes. This probability is often equated with the odds that a pot of water on a stove will freeze. Therefore, even if reversibility is assumed on the micro level, it does not have a noticeable effect on the macro level, since we would have to wait almost an infinite amount of time for the reversibility to make itself felt.

When quantum mechanics entered the picture, Boltzmann's application of Newtonian mechanics on the micro level had to be revised. Quantum theory does not see atoms as time-reversible billiard balls. The problem of time appears in the collapse of the wave function, or the so-called measurement problem. While it is true that the Schrödinger wave equation is time-reversible, the collapse of the wave function is not. The randomness encountered in quantum processes appears to be somewhat different from the randomness we normally encounter, in the shuffling of cards, for example. Before the wave function is collapsed, all the solutions of the wave function exist simultaneously. At this stage, the wave function is still reversible. After one solution is "selected" (or a measurement is made), there is no going back, even if one used an infinite amount of time — again pointing up the paradox of the reversibility of the Schrödinger equation and the irreversibility of the collapsed wave function. A solution was needed.

Quantum theory describes the possibilities for matter and energy. That is, the laws of nature set boundaries on what can happen. That combination of possibilities in itself leads to a symmetrical and time-reversible universe. But our three-dimensional world does not use all probabilities. Whenever anything happens, the symmetry is broken. John Briggs and David Peat in *Looking Glass Universe* explain it this way. Imagine the universe as a giant roulette wheel. The wheel is whole and symmetrical and contains an infinite number of slots. Each slot is a potential event. Once the ball falls into a particular slot, an event occurs — *one* event out of a multitude of probabilities. For anything to happen, symmetry must be broken. The timeless order of the wave function is changed into the timed order of the three-dimensional world where real experience takes place. This process is called the quantum mechanical arrow of time. Irreversible time is the result of breaking the symmetry of a timeless order.[54]

Prigogine suggests a resolution of the dilemma. For him, both times exist: the historical time of macroscopic structures and the reversible time of the underlying order. Even though the interaction of two particles seems to be time-reversible, in the real world this is an illusion. The interaction must take place in a larger complex system. Once this occurs — in the roulette wheel model, when the ball has fallen into a slot — time reversal is not just highly improbable (á la Boltzmann) but infinitely improbable. Looking at it from Bohm's point of view, the real event is enfolded in the timeless implicate order and unfolds into the explicate order, thereby creating time in our three-dimensional world.

Albert Einstein once commented on the variability of time by saying that an hour spent holding your girlfriend's hand seems like a second, but a second spent touching a hot stove seems like an hour. Because of clocks and other timing devices, our common experience is that physical time is absolute and does not depend on the circumstances. The special theory of relativity, on the other hand, tells us that physical time is variable and depends on the speed of the observer and indicates that we cannot separate the conception of space from that of time. Special relativity theory also states that physical laws are invariant for moving observers, but space and time are not. But if space and time are combined in the proper way, the changes in time due to relativity exactly cancel the changes in space, which results in an "invariant interval" of the combination.

In Bohm's view, since quantum theory says that matter is separated in space but really is a nonlocally related projection of a higher-dimensional reality, then moments in time can also be viewed as projections from this same reality. The implication is that time, like space, is a derivative of a higher-dimensional ground and is to be seen as a particular kind of order. Different time orders can be derived for different sets of sequences. The ground, however, cannot be comprehended in any particular time order or sets of orders. In short, time has an implicate order, and in any given moment the whole of time is enfolded.

According to Bohm, matter does not move across space-time as Descartes believed; rather it exhibits different levels of explication. The connections are not in space but in the implicate order. Time, then, is a succession, or an aspect of a succession, in the depth of implication. If we go far enough into the implicate order, what will later appear as succession is still co-present. Bohm quotes Nicholas of Cusa, a fifteenth-century mathematician and mystic, saying that eternity unfolds in time. Nothing really happens; everything just is. Nicholas of Cusa saw past and future always existing as overtones of the present. Time is a metaphor for the depth of implication. Bohm's conclusion is that there is fundamentally an immense multidimensional ground (holomovement) from which the projections determine the time orders.

In summation, let us consider the following analogy. Imagine a lake containing all possible events that can occur in our three-dimensional world, existing side by side and concurrently. The lake does not display movement as we understand it; that is, it does not move in space and time, but its movement can be seen as a flux. To use Bohm's terms, through this flux (or unfolding and enfolding), the universe is evolving to a higher unity. To use Seth's term, the lake grows in a qualitative way as an expression of its "value fulfillment." Let us identify the lake with the implicate order. Now imagine a stream flowing from the lake. By creating the stream, the lake "expresses" itself in a space-and-time order. The stream creates the timed order by stringing out co-present events in a sequential way. Events being fed into the stream are what we refer to as "projection" (unfolding). "Injection" (enfolding) refers to the return of events to the lake. Projection and injection occur in a pulsating manner.

The stream is the explicate order, where we live. The lake creates innumerable streams with a large variety of timed orders, and all this activity, as we shall see, is for the purpose of "growth" within the lake. The streams are suborders of time. The lake itself is beyond time.

SOMA-SIGNIFICANCE: MIND-BODY

One of the most vexing questions of philosophical thought is the so-called mind-body problem. The evidence is overwhelming that the mind affects the body. We also know that changes in the body affect the way we think and feel. While all this is accepted by almost everyone, the exact relationship between mind and body has long been controversial. Perhaps it is not an exaggeration to say that most scientists who think about this issue view the mind as an epiphenomenon of the body.

In 1949, Gilbert Ryle, an Oxford philosopher, wrote a book called *The Concept of Mind*. In it, he ridiculed the idea of mind by referring to it as the "ghost in the machine." His view reflected the outlook of orthodox science for at least the previous one hundred years. The successes of science during that period were apparently so heady as to preclude the need for any hypothesis other than that the universe consists of "matter in motion through empty space." But such a mechanistic view was not always in vogue. In Western thought, the idea of the mind as an entity different from the body goes all the way back to Plato.

According to Plato, the mind is entirely nonmaterial and has the capability of existing apart from the body. Furthermore, he felt that the mind directs the body and all of its activities. Plato's conception was not accepted by the atomists, represented by Leucippus and his disciple Democritus. They felt that all of nature consisted of hard little indivisible particles moving through space, and that any attributes of the mind were nothing more than the qualities and relations exhibited by this infinite number of particles.

In the seventeenth century Descartes developed a theory that was later labeled "Cartesian dualism." On the whole, Descartes tried to give all reality a mechanical and mathematical explanation. But he ran into difficulties. Starting with the premise, "I think, therefore I am," he had to introduce mental phenomena. He did so by assuming a second substance that had no relationship to matter. Matter was considered a physical, extended substance, and mind nonextended (i.e., not occupying space), indivisible, and immeasurable. But he could not describe a causal interaction between the two. When Newton and later mechanists returned to Democritus's little "billiard balls" traveling through space and affecting each other in numerous ways, Descartes' mind substance was left without a job.

The Platonic idea is that the mind is a separate entity that somehow affects the body, and the opposing view is that the mind is an epiphenomenon of the body. A third view was proposed by George Berkeley in the early part of the eighteenth century. In essence, Berkeley said that the body is an epiphenomenon of the mind. He felt no need to postulate any substance other than the spirit, or that which perceives. In his view, the body was left without a job. In the last few centuries, it has been an article of faith in science that the mental and physical are totally different — if the mental exists at all.

Bohm approaches the mind-body problem with his principle of soma-significance. Before we examine this principle, let us reiterate some of Bohm's basic ideas.

- Matter can be described as an explication of a deeper level called the implicate order.

- The explicate order comprises our three-dimensional world, but cannot be fully described without reference to the implicate order from which it springs.

- The implicate order is the source from which both our physical and mental worlds are created.

- The implicate order arises from an even deeper ground called the superimplicate order.

- There are deeper and deeper orders, all merging into the holomovement, which is the infinite-dimensional ground of All That Is.

- Each successive order unfolds and enfolds into the orders below and above.

While Bohm does not particularly like the word, he has described a "hierarchy" of levels wherein information from the higher level influences that of the lower level, while the higher level itself is influenced by the next level above it. So, in a sense, we have two aspects at each level, depending on our point of view. Looking downward, we see a more explicated order; looking upward, we see a more implicated order. Influences of information are passed upward and downward through projection (unfolding) and then injection (enfolding).

Each level is capable of organizing levels below into various structures or orders. We have an explicate order, which receives its "guidance" from the implicate order, which in turn receives its capability for organizing structures from the superimplicate order. (This concept of a continuum of ordering principles is similar to Bohm's ideas described earlier regarding a spectrum of orders. In that notion, the context of the random order must be expanded to find order in randomness. This spectrum of order is infinite; a level of total determinism is never reached.)

Bohm generalized these views into his principles of soma-significance and signa-somatic. He avoids the term psycho-somatic because it suggests two different entities in some sort of mutual interaction. He prefers the concept of soma-significance, which portrays soma (physical) and significance (mental) as two aspects of one overall reality. Each aspect — defined as "a view or a way of looking" — reflects and implies the other.

Physicist Wolfgang Pauli points to a similar unity of the physical and the psychical:

On the one hand, the idea of complementarity in modern physics has demonstrated to us, in a new kind of synthesis, that the contradiction in the applications of old contrasting conceptions (such as particle and wave) is only apparent; on the other hand, the employability of old alchemical ideas in the psychology of Jung points to a deeper unity of psychical and physical occurrences. To us . . . the only acceptable point of view appears to be the one that recognizes *both* sides of reality — the quantitative and the qualitative, the physical and the psychical — as compatible with each other, and can embrace them simultaneously. . . . It would be most satisfactory of all if physics and psyche could be seen as complementary aspects of the same reality.[55]

Each aspect, soma and significance, is displayed or unfolded either in perception or in thinking. The separation of the two occurs only in our thoughts. The relationship of soma and significance is such that each type of significance is based on the corresponding structure of soma. This relationship is carried out through what Bohm calls "meaning." Meaning as used by Bohm is difficult to define since it is capable of indefinite extension. However, I believe it is fair to say that meaning is a form of being, and being is that which has actuality (material or nonmaterial). In this sense, meaning is the essential feature of consciousness, and matter and everything else is permeated with meaning. From this, one can conclude that an elementary particle has meaning, but is not necessarily self-aware.

Meaning, for Bohm, is the bridge between soma and significance. It is the activity of information and is the essential feature of consciousness. The metaphor of a bridge to describe meaning is meant to indicate that the physical and mental are inseparately linked. Meaning is present in both aspects. The bridge carries information, which gives rise to activity. The activity can be actual or virtual; that is, it can be a group of possibilities of which only one becomes actual. The activity, then, is meaning that can (for example) reorganize the material aspect.

Based on Bohm's model of the causal interpretation of quantum mechanics on the atomic level, the electron (soma) has a field (significance) around it that obeys Schrödinger's equation or the resulting wave function. The wave function is the mental or significance aspect of the electron. The field (wave function) and particle are never separate and are actually aspects of the same reality. The field acts on the particle, not by its intensity, but by its form (information). This information, or form, or field, gives rise to an activity that is identified with meaning. The meaning (arising out of the field, and, in this case, a kind of protointelligence) guides the electron as a radio wave guides a ship.

The information field, which is the significance aspect of the electron-field relationship, is multidimensional and is enfolded in a multidimensional ground. It gives rise to an activity of meaning that determines the interrelationship of the particles. The wave function, which determines the field, is analogous to a musical score for a ballet or a dance. The dance is the meaning or activity of the information or score. The wave function (score) is not fixed, but depends on the initial configuration of the particle, just as the dance depends on the initial position of the dancer.

Bohm uses this same concept on the macro level to model our activity when we react to a situation. Suppose we see something on a dark night. This image is unfolded in our perception as either an assailant or a harmless shadow. If the meaning is "assailant," the blood pressure rises and a rush of adrenalin takes place (soma). The meaning acts upon matter to produce a certain behavior. If the image indicates a meaning of "harmless shadow," then the soma reacts in a different way.

Bohm extends the idea of meaning to include a multilevel arrangement in the holomovement. Meaning always depends on the context, so it can never be fully defined. To establish this multilevel arrangement, Bohm introduces an additional aspect of reality that he calls subtlety and which he describes with such terms as rarefied, delicate, and intangible. Subtlety is contrasted with what is manifest. The distinction between the two is relative, so that what is manifest on one level can be subtle on the next.

Thus, soma-significance has degrees of subtlety depending on the level. Bohm gives the example of a written page in a book. Somatically, it is ink on paper with a certain form. The significance or content of the page is more subtle than the somatic aspect. Taking it a step further, the significance of the page can be held in another somatic form, such as the human brain, with all its attendant chemical and electrical processes. This somatic form of the brain is more subtle than the somatic form of the book which gave rise to the significance.

The thought, which is the significance of the somatic brain, may have a meaning that can be grasped by even higher and more subtle somatic processes. This brings about meanings of higher and higher levels.

Perhaps it is advisable to state again that soma and significance are aspects of one totality. Since the hierarchy is one of subtlety, looking downward, a soma aspect is exhibited; looking upward, a significance aspect is exhibited. Both are the same "substance," different only in subtlety. What is seen as soma on one level can be considered significance from a lower, less subtle, level. The meanings at the various levels, acting as a totality, can be the origin of a flash of insight.

Meanings can be enfolded in each other and in an implicate order of indefinite extension. Insights arise from the depth of implication of the meaning. The depth of implication is correlated with the degree of subtlety. The meanings at the various levels are the activities of the various fields (significance) at those levels. In other words, the organizing field identified with the particle may in turn be organized by superorganizing fields, which in turn are organized by additional fields, shading off into an infinite number of implicate orders, superimplicate orders, super-superimplicate orders, and so forth. Meaning is discussed more fully in Chapter 4; suffice it to say at this point that meaning is an unending spectrum of consciousness. In Bohm's words,

> A change of meaning is a change of being. If we say consciousness is [the] content [of being], therefore consciousness is meaning. We could widen this to a more general kind of meaning that may be the essence of all matter as meaning.[56]

Human consciousness is only one portion of that spectrum (as visible light is only one portion of the electromagnetic spectrum). Matter, which is on the "lower" end of the spectrum, is also consciousness, or a kind of protointelligence. Although meaning is a form of consciousness, it is not necessarily self-aware, as in human consciousness.

Before we proceed with the reverse process, signa-somatic, we need to clarify a bit further Bohm's views as they apply to the quantum world. To do that, we must understand the concept of the formative cause. Bohm defines the formative cause as that which gives form to the activities of an entity and provides a goal toward which the entity is moving. Given this definition, the formative cause is similar to meaning. In the case of the sentient being, its formative cause is its consciousness. Bohm sees the wave function as a partial description of the formative field of the electron. The formative field has a type of meaning, or consciousness, that provides for such concepts as nonlocal correlations. The individual electron has its own wave function; the total system also is represented by its own wave function. In this way, consciousness can be seen as connected with the total system as well as with its parts.

In an interview with Renée Weber, Bohm was asked if there is meaning in the nonhuman world and in the universe as a whole. He replied that not only is there meaning to it, but that it *is* meaning. Would this apply to the subatomic world and the cosmos? Bohm says yes. In the case of the electron, Bohm points out that the term "meaning" does not imply that it is self-aware in the human sense. The degree of self-awareness in the particle world is quite low, but is not absent. In Bohm's view, the wave function is not just a description of a material object but is multidimensional and cannot be placed in our three-dimensional world. This implies that space and time also have meaning in some sense.

The process of organizing activity on more subtle levels can be reversed. The illustration we used earlier of encountering a shadow is an example of the reverse process. The significance of the shadow as "assailant" creates a meaning in the soma, causing the blood pressure to rise and the adrenalin to flow. Such bodily changes indicate that the significance on one level can actively affect the soma on a more manifest level. Bohm calls this reverse process signa-somatic. Literally, then, any change in meaning is a change of soma, and a change of soma is a change in meaning. A two-way movement of energy takes place in which significance acts on the more manifest soma, and perception (through display) carries the meaning back in the other direction.

Two points need to be emphasized. First, the levels we have been discussing do not represent distinct stages; they are abstractions only. Second, we must remember that in this to-and-fro process nothing exists except the flow of information or energy (not through space-time). The meaning is carried between the soma and significance on a given level and between levels that are either more or less subtle or manifest.

The soma-significance and signa-somatic processes can be extended indefinitely. As an example, meaning is conveyed between members of a society by somatic means such as sound waves, books, telephones, radio, and television. These forms of communication convey meaning. Material objects, such as buildings that we erect, are somatic results of the meaning that such objects have in the "significance" of human beings. (These observations are reinforced by some of the Seth material, e.g., see p. 128.)

In terms of the implicate order, meaning (or consciousness) is constantly unfolding and enfolding, and, in the process, actualizing structure. But meaning is never complete. Since the depth of these hypothetical levels is infinite, each meaning enfolds more comprehensive meanings without end. As we create discrepancies on a particular level, these are clarified by a more comprehensive meaning at a deeper level. But we will always encounter new discrepancies. Knowledge will always be incomplete.

There is movement also from the explicate to the implicate, so new meanings are being created from moment to moment. If this were not the case, significance would come solely from memory, and we would be limited to a finite depth of implication. To increase our level of comprehension, we must have access to unlimited depths of implication. This access to greater depths allows us, in effect, to create new meanings in our memory.

Since we have access to infinite depths of implication, at each deep level we encounter a certain degree of ambiguity because the next deeper level has not been encompassed. As Bohm himself asks, is there a bottom level of reality where ambiguity is not encountered? Where the meaning would be clear no matter who looked at it, and independent of the meaning we might bring to it? In answer to these questions, Bohm comes to the same conclusion as Bohr in his Copenhagen interpretation of quantum theory, but for different reasons. That is, there is no bottom level without ambiguity.

At this point, let us discuss the terms *context* and *content*. Content is the essential meaning abstracted from a wider context, and therefore the meaning is not fully defined without taking the context into account. This context-dependence is illustrated by the need for indefinitely higher and more subtle levels of meaning (consciousness) to clear up ambiguities. The process is infinite, and the ambiguities are never totally eliminated. The problem of ambiguity is also present in physics when matter is under consideration.

According to quantum theory and the Copenhagen interpretation, matter is context-dependent. Bohr found that quantum theory contradicted classical theory, but required the use of the same classical concepts that were being contradicted. To get around this paradox, he simply stated that the quantum theory introduces no basically new concepts. He took the concepts of position and momentum, which were unambiguous in classical theory, and made them ambiguous in quantum theory. Since quantum theory fails to predict single events and can give only statistical predictions, it cannot provide an unambiguous picture of the process taking place. Bohr said that in order to get meaning out of process, the whole context of the experimental arrangement must be included. The results depend on large-scale behavior. Therefore, it is not possible to have a "bottom level" that is unambiguous.

Bohm agrees with Bohr's conclusion but comes to it in terms of the implicate order. In the concept of the implicate order, reality is not mechanistic but is understood through the process of enfoldment and unfoldment. The total implicate order allows for sub-wholes, or elements, to be abstracted, but only in a relatively independent way. The activity of each element is context-dependent on the whole; the larger content (significance) organizes the sub-wholes into one greater, more inclusive whole. The net result of all this is that matter and mind are context-dependent and can be viewed as processes of enfoldment and unfoldment between levels of subtlety. As a result, there can be no bottom level where ambiguity is eliminated.

According to Bohm, everything has a physical and a mental aspect. In inanimate matter, the mental aspect is very small. As we go deeper into the implicate order, the mental aspect becomes more and more important. Once again, there is no unambiguous bottom level such as one might postulate for a machine. Nature is infinite in its potential depth of subtlety.

The same element can display both the soma aspect and the significance aspect, depending on the context. Physicist Heinz Pagels relates an interesting story that is particularly useful in suggesting this interrelationship.

Once a couple of engineers wanted to make a computer model of a physical process involving some aerodynamic design problem. They decided to build a simple computer that was especially designed to computationally model these physical processes. While they were working out the design of the computer a friend came by. He announced that they were wasting their time. Instead of building a small special-purpose computer, he said, they should model their special-purpose computer on a more powerful supercomputer that was available. Then they could run their original computer "inside" the supercomputer. This was done. What is interesting is that what was originally a hardware device became software within the larger computer. Maybe (and I think mind-monists would concur) all the hardware we see around us — the material world — is like the little computer inside the supercomputer, really just software, a representation of information. Everything is mind. One could, of course, say with equal satisfaction (as materialist-monists would) that everything is really matter, even consciousness.[57]

This illustrates how the special-purpose computer can be viewed as either soma or significance, and demonstrates that the material world can be seen as a representation of information. Whether reality is all mind or all matter is a question we will return to throughout this book.

THREE-DIMENSIONAL WORLD
AS A MULTIDIMENSIONAL DERIVATIVE

The current cosmological model for the origin of the universe is called the big bang theory. Briefly, it states that all matter and radiation were created at a finite time in the past (10-20 billion years ago). This theory has been very successful in explaining such phenomena as the expansion of the universe and the microwave background radiation. In the big bang view, the universe started out from an initial state of extreme temperature and density and then began expanding and cooling.

One of the metaphysical problems with the big bang theory is that whatever existed before the big bang is an emptiness that stretches the imagination, to say the least. It must be a domain that cannot be defined by space, time, matter, and energy. But it did contain something — and that something is the principle guiding the laws of physics.

According to the big bang theory, this ultimate nothingness was fertile with physical law. If we accept this premise, then law existed in a transcendental realm, since space-time did not yet exist. Therefore, the laws of nature, although they govern the physical universe, cannot be considered physical.

Bohm sees the big bang as just a little ripple on the vast sea of energy of the holomovement. The ripple is regarded as a sudden wave pulse from which the universe began. The pulse then exploded and separated out, thereby creating our expanding universe. The implicate order would have had space and time enfolded, which was then explicated. That is, the space-time continuum existed in potential form in the implicate order and was unfolded to become our three-dimensional universe. This insight convinced Bohm that the universe cannot be independent of the sea of energy.

Furthermore, Bohm views the fundamental activity of existence as light (or a similar form). The holomovement is described as an unending sea of light; matter is condensed or frozen light. This condensed light (matter) is not made up just of the visible portion of the electromagnetic spectrum; we are speaking here of the entire range of waves that travel at the velocity of light. The patterns forming matter are rays of electromagnetic radiation that are somehow reflected back and forth and thus become stable. If the reflection stops, then matter turns into energy. Physicists at the present have no theory that explains the reflection process.

To return to the ripple, at a certain stage of an eternal universe of light, some light rays combined and produced the big bang. The universe of the big bang will probably end, but the universe of light is beyond time and has no ending. One can also assume that there are other universes, and other big bangs, all emanating from the eternal universe of light.

Originally, there were no particles, no space, no time — although, according to the big bang theory, all these developed in a very short time indeed. Bohm points out that physicists assume that the laws of quantum mechanics were in force before and after the creation of matter. But since quantum mechanics is based on the results of measurement, before mass and space came into existence, there was nothing to measure. Seen in the context of the implicate order, the universe before the big bang can be thought of as relatively unformed energy. Apparently, formative fields gradually became necessary, and at a certain moment, these fields organized the amorphous energy into the original particles.

The formative fields are not eternal but are constantly forming and developing. So the entire structure of quantum mechanics arose from some other state, in existence before space and time. Dealing with the formative field before space and time is not in the purview of physics as presently constituted. This level of meaning in the implicate order can be reached, so far as we know, only by going inward and experiencing it directly — the way of the mystic.

FORM, A PROCESS OF PROJECTION AND INJECTION

The processes of explication involve what Bohm calls projection and injection in the creation of form. The number of forms in the universe, and their extraordinary diversity, is tremendous, and goes far beyond what is necesssary for survival. Here we are defining form as the structure or pattern in the universe, the shape and organization of a system, as distinguished from matter itself.

In the processes of projection, injection, and then reprojection, the first projection does not directly cause the second projection. Rather, it is through the injection phase that the first projection influences the second projection. Since the influencing does not take place through the explicate order, it is nonlocal and takes place in the implicate order. Influences everywhere can mediate the reprojection. This interpretation effectively rejects the mechanistic view with its emphasis on the external relations of entities. The implicate order is a common ground to all forms of matter, and these forms are related *internally* rather than externally through causal means. All causation occurs through the process of projection and injection.

The mode of projection and injection of moments (a moment is one complete cycle of projection and injection) creates a constant component in which past forms are replicated in the present. Since the implicate order is not local, similar forms resonate and are connected regardless of their location in space and time. Through the process of resonance, particular forms are reinforced and appear in the manifest world as one stable form. The stable form can appear anywhere, and all similar forms influence each other in the implicate order (or the unmanifest). The implicate order then is the source for all new forms that are explicated. Bohm does not feel that forms pre-exist in the Platonic sense, but are developed and modified through the processes of projection, injection, and reprojection. What we have is a creative implicate order being guided in some sense by the explicate order through the process of injection, a kind of feedback system.

Weber points out in *Dialogues With Scientists And Sages* that this view is in contrast to traditional systems in which God directs operations from above. In Bohm's system, the whole acts together as a unit. We all contribute in some way through the display given us by the explication process. The implicate order cannot function without the explicate order. Such an interchange implies that we are part of an eternal life-energy, a part that can act only when the explicate is displayed to it. The whole point of the manifest world is to relatively separate consciousness and matter, whereas they are not separated in the implicate order. The two elements can then interact in the manifest world to satisfy the purpose and intent of the implicate order. As Comfort so succinctly phrased it, "Explication postulates an audience."[58] And Bohm sums it up: "Life-energy is more than just biological organization; it reaches into intelligence. Guided by a display, it can do almost anything, but without a display it has nothing to do."[59]

Projection and injection allow us to live in a somewhat stable explicated world, and yet they provide us an avenue for creativity. Maintaining the balance between creativity and stability is vital to functioning in our manifest world. If all explications were creative, everything would disappear as soon as it appears. There would be no past. On the other hand, without creativity, we would be stuck in our past with no hope for release. All events would be rigidly determined. The universe seems to be experimenting and thereby learning. Forms that do not last seemingly are not compatible with what we might consider the purpose of the universe. The implicate order, with guidance from the explicate, constantly creates new forms that either resonate, and thereby survive for a time, or disappear. A process of this kind indicates purpose for our universe.

WHOLENESS AS AN ASPECT OF
QUANTUM AND RELATIVITY THEORIES

Throughout Bohm's discussions of the causal interpretation of quantum theory and its expansion into the concept of the implicate order, one basic feature is paramount: his commitment to wholeness. That sense of wholeness is unswerving and fundamental to all of Bohm's ideas. Even in the early 1950s, he noted that the wave function is not in one-to-one correspondence with the actual behavior of matter. One of two observables that do not commute[60] does not have a definite value if the alternate observable is being measured. This fact suggested to Bohm that in some sense quantum theory was less mathematical in its philosophical basis than classical theory. The universe is not constructed according to a precisely defined mathematical plan. The wave function can be seen as an abstraction that provides a mathematical reflection of only some aspects of reality but does not give a one-to-one mapping. The mathematical correspondence is statistical, not precise. This implies that the universe cannot be made up of separate elements having corresponding mathematical quantities in the theory. Rather the universe must be seen as a whole, from which we extract aspects to examine. In short, there are no separately existing and precisely defined elements.

The necessity for wholeness shows itself in numerous other ways as the reality beneath quantum theory is examined. For one thing, if a hidden variable theory is possible at all, Bell's theorem (along with physicist Alain Aspect's experimental results) shows that nonlocality is an absolute requirement. Herbert says that to be consistent with known facts and Bell's proof, a universe that displays local phenomena must be built on nonlocal reality. With this stringent requirement in mind, Bohm sees a necessity for wholeness when he examines the two main theories governing modern physics, quantum theory and relativity theory. He states:

> Relativity theory requires continuity, strict causality (or determinism) and locality. . . . Quantum theory requires non-continuity, noncausality and non-locality. So the basic concepts of relativity and quantum theory directly contradict each other. . . . What is very probably needed instead is a qualitatively new theory from which

both relativity and quantum theory are to be derived as abstractions, approximations and limiting cases. . . . The best place to begin is with what they have basically in common. This is undivided wholeness. Though each comes to wholeness in a different way, it is clear that it is this to which they are both fundamentally pointing.[61]

While this statement appears convincing regarding the need for any superquantum reality to be nonlocal, the idea of nonlocality sends shudders through the physics community. To most physicists, nonlocality means instantaneous communication, and this stabs at the very heart of relativity theory. More important, it leaves the door ajar for the entrance of numerous time paradoxes difficult to handle on a logical basis. Einstein himself found the idea of nonlocality totally unacceptable and was never comfortable with quantum theory in spite of its tremendous success. In a famous statement he said, "Quantum theory cannot be reconciled with the idea that physics should represent a reality in time and space, free from spooky action at a distance."[62]

Nonlocality is at the center of Bohm's theory. Bell states this directly:

I thought that the theory of Bohm . . . was in all ways equivalent to quantum mechanics for experimental purposes, but nevertheless was realistic and unambiguous. But it did have the remarkable feature of action-at-a-distance. You could see in the equations of that theory that when something happened at one point there were consequences immediately over the whole of space unrestricted by the velocity of light.[63]

What is said is that these consequences are sent across space and time without passing through the space in between. This is essentially the definition of nonlocality if relativity is to be satisfied.

Bohm feels that causal paradoxes do not present a basic problem, mainly because the influences are not propagated through space and time but through the absolute space-time in the background that underlies the quantum. As Hiley (his associate) points out, the orthodox interpretation of quantum theory does not allow the physicist to ask the question, what happens between two systems separated in space? All the physicist has is a wave function with which to work out the correlations. But the wave function does not offer any explanation of what goes on underneath. Hiley says:

The quantum potential will not have any causal paradoxes in it because it essentially requires an absolute spacetime in the background . . . . We take the field theory and construct a superpotential from the fields. You can show that the superpotential (which is governed by a Schrödinger superwave equation) is in instantaneous contact with all particles (i.e., nonlocal). But when you work out the statistical results of typical quantum experiments you find that . . . they obey the theory of relativity. So in other words, relativity in the quantum potential approach comes out as a statistical effect, not as an absolute effect.[64]

Bohm views nonlocality as arising from a new kind of coordinated movement. He says that the nonlocality is not the result of "pre-established harmony," not the result of a "conspiracy." Instead Bohm sees it as a stable form of the overall movement, carried in a continuous and local way by the "vacuum" fields in the background. In other words, the result of measurements carried out in one place is not totally independent of measurements carried out in a second place.

This concept offers a possible explanation of a thought experiment by Einstein, Podolsky, and Rosen (the EPR experiment), which attempted to show that quantum theory was not complete and that a set of hidden variables was needed (see Chapter 8). The idea behind Bohm's wholeness is that we cannot separate particles as postulated in the EPR experiment. Particles cannot be treated as independent units; rather they should be envisioned as aspects of the whole. As Hiley has said, it is the whole system that responds.

Electrons do not have cogwheels or computer parts as in the discarded mechanistic approach; we recognize their interdependence from the correlations that we see. It is because of our limited vision that the particles appear to be independent. The closer we get to the underlying world of nonlocality, the more we run into difficulty in breaking up the whole into arbitrary (and nonexistent) particles. We can separate objects only when we deal with relatively large numbers of particles. Reality must be basically whole.

SUMMARY

Quantum mechanics tells us that our macroscopic world rests on a microscopic world that is ambiguous, ruled by probabilities — that God does indeed play dice with the universe. We can either accept the

fact that the underlying reality is random and live with the philosophy of positivism, or assume that there is a more basic reality beyond
the quantum level. Following Bell and Aspect, whatever that subworld
is, it must have the attribute of nonlocality; reality has to be interconnected. At the very least, the basic particles (if such exist) must not be
inert "billiard balls" but aspects of the whole. If relativity theory is
correct, then that connectedness must exist outside our three-dimensional world.

Bohm's ideas rest squarely on a model that postulates undivided
wholeness as the ground of reality. He suggests that both relativity
theory and quantum theory convey the message that everything in
the universe merges and unites into one total reality, which he defines as the holomovement. The basic quality of the holomovement is
enfolding and unfolding. Through this process, matter is formed out
of the totality. The holomovement is made up of, or carries, an unending spectrum of levels of order, which Bohm designates as implicate
orders. The implicate order concept is a metaphor for wholeness. Each
level may appear random or chaotic, but when imbedded in a larger
context, the randomness gives way to order. Hidden in the apparent
random behavior is a manifest order that can be partially explicated
into things and thoughts. The things and thoughts are manifest aspects of the implicate order from which they came and are called the
explicate order. So, in the explicate order we have a separation of
objects; in the implicate order we have interconnection, wholeness,
and everything enfolded into everything else.

The age-old mind-body problem is examined by Bohm in terms of
the holomovement. At any given level of the holomovement, there
exists an explicated aspect that he calls soma and an implicated order that he calls significance, which gives rise to the soma. That is,
each level of the totality has a material side (soma, or explicated) and
a mental side (significance, or implicated). The relationship between
the two sides of a particular level is mediated by what Bohm calls
meaning (or consciousness). Meaning has the ability to use information to guide the soma. This information comes from succeeding levels of order. At every level of the holomovement, the soma aspect is
more manifest in the explicate direction and more subtle in the implicate direction. The spectrum of order can be viewed as a spectrum of

subtlety; higher degrees of consciousness (or meaning) are associated with deeper and more subtle levels. Insight and creativity find their sources in the depths of the holomovement.

Assuming Bohm's view of reality, within the ambiguity and seeming chaos of the world beyond the quantum lies a new level of order. It will be described eventually by a new theory that has enfolded in it both quantum theory and relativity theory, just as Newtonian mechanics is enfolded in those theories from the twentieth century. The notion of an enfolded or implicate order seems destined to be an important concept in any new paradigm.

2

The

Perennial Philosophy:

A Mystical Perspective

Relativity and, even more important, quantum mechanics have strongly suggested (though not proved) that the world cannot be analyzed into separate and independently existing parts. Moreover, each part somehow involves all the others: contains them or enfolds them. . . . This fact suggests that the sphere of ordinary material life and the sphere of mystical experience have a certain shared order and that this will allow a fruitful relationship between them.

David Bohm

*T*he term "Philosophia Perennis" was introduced in its current sense by philosopher Godfrey Leibniz (1646-1716) and made famous by Aldous Huxley with his book *The Perennial Philosophy*, first published in 1944. In this chapter we deal with the concept of mysticism as exemplified by the Perennial Philosophy. Our treatment focuses on those aspects of mysticism that relate to the physical theories discussed in Chapter 1.

THE PERENNIAL PHILOSOPHY: THE MYSTIC'S METAPHOR

Mysticism is an expression of an inner wisdom reached through an altered state of consciousness. This altered state is analogous to the implicate order in that it aligns one with the wholeness of reality. Esoteric mystical traditions are present in all the major religions of the world; even the more primitive religions include simple

versions of such traditions. In fact, many religions are based on stories of particular mystical experiences. These stories gave rise to the
many sets of theological beliefs found throughout the world. The established faiths attempt to place the report of the mystical experience
in understandable terms; hence, ritual and liturgy, pageantry and
spectacle, sacred books and writings constitute the "container," which
is determined by the cultural environment. Doctrine takes its specific
form from this container. In this way, the other-worldly experience is
made accessible to the religion's followers.

While the established religions are obviously quite different in
their superficial dogmas, Huxley found a common core in all their
theologies — a shared implicate order, so to speak. He called this the
Perennial Philosophy, defined as the transcendental essence of all
the main religions promulgated through their mystical traditions. The
metaphysics of the Perennial Philosophy, Huxley says, is "immemo-
rial and universal."

Erwin Schrödinger, who was interested in mysticism, commented
on Huxley's Perennial Philosophy:

> [It] is an anthology from the mystics of the most various periods
> and the most various peoples. Open it where you will and you will
> find many beautiful utterances of a similar kind. You are struck by
> the miraculous agreement between humans of different race, dif-
> ferent religion knowing nothing about each other's existence, sepa-
> rated by centuries and millennia, and by the greatest distances that
> there are on our globe.[1]

While these transcendental traditions have obvious similarities,
we should not assume that there is absolute agreement among them.
And while the Perennial Philosophy is sometimes equated with the
mystical experience from which it springs, each tradition has its own
special insights and beliefs, and each individual's encounter with the
unknowable is unique.

The mystic who has a direct experience of the divine Ground (or
the mystery) cannot have that experience invalidated or validated by
any of the sciences, or, for that matter, by any traditional religion. As
soon as he/she describes the experience, however, a version of the
Perennial Philosophy is created. The spoken or written description of
the experience is a metaphor or a "map" of the territory, and, as a

mental construct, is open to a rational critique. The difficulty lies in the fact that the mystic is confined to a language geared to a level of reality different from that of the experience. Huxley himself states this very well when he notes that a truly correct expression of the Perennial Philosophy is not possible. He says:

> Nobody has yet invented a Spiritual Calculus in terms of which we may talk coherently about the divine Ground and of the world conceived as its manifestation. . . .

> So far, then, as a fully adequate expression of the Perennial Philosophy is concerned, there exists a problem in semantics that is finally insoluble.[2]

Nevertheless, the body of mystical traditions contains remarkably similar descriptions of a transcendental unity. In all the religions the Perennial Philosophy describes an Absolute Ground, which is the reality of all things. The Absolute is not set apart as some sort of Creator separated from that which is created. Rather, as the ground of All, the Absolute is one, completely whole and indivisible. There is a tendency in the West to identify this wholeness with the philosophy of pantheism, but that is incorrect. The Absolute is not just *in* all things, all things *are* the Absolute. Nor is the Absolute limited by our conception of the universe: the universe is only one manifestation of an ultimately undifferentiated and indefinable reality.

As stated by Huxley, the basic premise of the Perennial Philosophy is that the eternal self[3] is one with the Absolute, and each individual is on a journey to discover that fact. Each individual's path through life is to fulfill his/her destiny and thereby to return to his/her true home. This is accomplished through remembering our true nature rather than through learning.

In examining the Perennial Philosphy, we will rely primarily on the writings of Ken Wilber. A transpersonal psychologist who received his training at the University of Nebraska, Wilber has extended his study beyond psychology to other disciplines.[4] Wilber defines transpersonal psychology as "a sustained and experimental inquiry into spiritual, or transcendental, or perennial philosophical concerns." His work presents transpersonal psychology as the modern expression of the Perennial Philosophy, or the application of the Perennial Philosophy to psychology.

The transpersonal psychology of today has its roots in the psychological thought of the past. The idea of the unconscious was introduced by Sigmund Freud and developed and refined by Carl Jung. To Freud, the unconscious was the warehouse for repressed drives, and the purpose of psychoanalysis was to bring these drives to the conscious mind. Jung, on the other hand, saw the unconscious as both personal and connected to the vast collective unconscious of all humanity. To him the collective unconcious was transcendental and expressed itself in universal symbols — Jung's famous archetypes.

Jung's theories on the genesis and development of our universe include a ground from which all reality springs. He referred to this ground as the *pleroma*, an old Gnostic term for the potential from which physical nature arises. One of its features is that each point in the pleroma, no matter how small, contains the whole, which is infinite. This, of course, reminds us of the hologram and the implicate order. Furthermore, that which is formed from this infinite ground is called the *creatura*. The creatura is like our three-dimensional universe, or the explicate order. The Jungian archetypes are formative elements that originate in the pleroma, which Jung broadens to include the nonphysical realm of the collective unconscious. The archetypes are patterns or symbols available to all individuals from all cultures, past and present. The unconscious is the ocean that gives birth to the individual "wave packets"[5] of consciousness; the wave packets are the conscious ego. The ocean contains the whole self; the whole self exists as potential for the ego.

Jung's description of the self as transcending the ego (personality) gave the field of transpersonal psychology an impressive start. He saw the self as transcendental, the directing force, on a higher level than the ego, and surpassed Freudian concepts with his idea that human reality springs from much deeper ground.

Wilber notes that all formulations of the Perennial Philosophy posit a hierarchic chain of consciousness in which the self journeys from the lowest, most dense, and fragmentary level to the highest level, which is the source and nature of all the levels. We shall be concerned with the shared features of this chain, especially the lower levels, where it finds a common ground with physics. Also, as we shall see, the self, the superquantum potential, and Seth's inner and outer ego are all analogous elements of this hierarchy.

CONSCIOUSNESS AS A SPECTRUM
OF INTERPENETRATING LEVELS

The basic feature of the Perennial Philosophy is that consciousness is displayed as a hierarchy of dimensional levels. In 1802, William Blake, the English poet and mystic, described his view of this in the following verse:

> Now I a fourfold vision see
> And a fourfold vision is given to me
> Tis fourfold in my supreme delight
> And threefold in soft Beulah's night
> And twofold always. May God us keep
> From single vision and Newton's sleep.

As expressed through his sometimes complex symbolism, Blake saw consciousness in levels. The number of levels is not significant; the interesting point is that the lower level is only an aspect of, and is subordinate to, the level directly above. Bohm would say that there exists a continuum of ordering principles. The simple vision is the lowest grade and is equated with the Newtonian outlook; that is, the lowest level is exactly what we know through our five senses — no more, no less. Thus consciousness progresses upward toward the level of the mystic, which Blake calls the fourfold vision.

The lowest level of consciousness is the most dense and the most fragmentary. Wilber labels this level "insentient consciousness." The higher we go, the less dense and more wholistic the levels become. Density is used here in a manner similar to, and inversely related to, Bohm's concept of subtlety. It is important to keep in mind that the levels are not separated, not discrete, but mutually interpenetrating and interconnected. Wilber quotes A. P. Shepherd:

> These "worlds" [or dimensional levels] are not separate regions, spatially divided from one another, so that it would be necessary to move in space in order to pass from one another. The higher worlds completely interpenetrate the lower worlds, which are fashioned and sustained by their activities.[6]

What divides or specifies the levels of consciousness is focus. The lower levels represent a more limited focus than the levels above. Because of this limitation, the consciousness on the lower planes cannot experience, or is not even aware of, the world above it. This is true even though the higher worlds interpenetrate and sustain the levels below. On the other hand, a consciousness can move up to a higher level just by broadening its focus. Then the higher world becomes manifest, and the consciousness exists on a new plane.

Although all levels of consciousness are available (with the proper vision to perceive them), individuals tend to create boundaries of existence that limit them to focusing on one level only and those levels below it. Huston Smith uses a metaphor in his book *Beyond The Post-Modern Mind* that is applicable here. He says that the divisions between the stages of consciousness are like one-way mirrors. If we look up, we see only the reflection of the level we now occupy. Looking down, the mirrors are as transparent as plate glass. When the self reaches the highest plane, even the glass is removed, and pure interpenetration exists. If, as Huxley says, the eternal self is on a journey, that journey can be seen as a process that widens the boundaries of comprehension.

The hierarchic chain of consciousness has different numbers of levels, depending on the version of the Perennial Philosophy. There may be two levels (matter and spirit), three levels (matter, mind, and spirit), or dozens of levels. The number is less important than the fact that consciousness is displayed as a hierarchy. Wilber describes six levels:

1. Physical: nonliving matter/energy
2. Biological: living (sentient) matter/energy
3. Mental: ego, logic, thinking
4. Subtle: archetypal, trans-individual, intuitive
5. Causal: formless radiance, perfect transcendence
6. Ultimate: consciousness as such, the source and nature of all other levels.[7]

From a general point of view, each level transcends, but also includes, all lower levels. Because the higher level transcends the lower, the higher cannot be derived from or explained by the lower. And while a higher level contains the attributes of the lower level, it also has new aspects clearly different from those of the lower and cannot be seen as a derivation of the lower plane. This notion is in contradistinction to the reductionism of nineteenth-century physics.

Wilber uses the metaphor of a ladder to describe the hierarchic structure. From each rung the view of reality is broadened, more inclusive, more unified, and more complex than from the previous rung. To use a dimensional analogy, the three-dimensional sphere contains the two-dimensional circle, but not vice-versa. This idea is similar to Bohm's concept of the spectrum of order. Climbing Wilber's ladder broadens the context. In Bohm's construct, what seems to be random on one rung becomes more ordered on a higher rung.

Wilber subdivides each level of consciousness into a "deep structure" and a "surface structure." The deep structure contains all the potentials of that level along with all its limits. In essence, the deep structure is a paradigm, and as such it contains the whole set of forms for that level.[8] The limiting principles within the deep structure determine which surface structures are actualized. Wilber defines the ground unconscious as all the deep structures of all the levels existing as potentials ready to emerge. So, in a sense, each deep structure contains the potential of all deep structures in an enfolded order. All the deep structures of all levels might be considered a counterpart of Bohm's holomovement.

The surface structure of a level of consciousness is a particular manifestation of the deep structure. While the surface structure is constrained by the paradigm of its deep structure, the surface structure can manifest or unfold any potential within the limits of that deep structure. In *The Sociable God*, Wilber compares the relationship between the surface and deep structures to a game of chess. The surface structures are the various pieces and the moves they make during the course of the game. The deep structures are the rules of the game. The deep structure, via the "rules," wholistically unites each piece to all the others. Each game played has a different surface structure, but all games share the same basic deep structure. Thus, we can say that the deep structure is the nonmanifest order; it does not have a separation in space or time. At least on the lowest, material level, the surface structure is the explicated (manifest) portion — our material world. The deep structure is interpenetrating and interconnected, but the surface structure is relatively separated, with a space-time attribute.

Wilber defines any movement of the surface structure as translation.

Again using the chess game as an analogy, translations are moves in the game. As for what causes the moves in the game, Wilber says that they are both (a) the sum of previous moves in the game thus far, and (b) the judgment of the player in reference to previous moves. He indicates that translations are historically conditioned but not totally determined. Somehow, the judgment of the player is a factor. There seems to be a mixture of causal and creative components. Furthermore, a particular level maintains itself by a series of more or less constant translations.

The movement from one deep structure to another Wilber calls transformation. What is it that is doing the moving from one level to the other? None other than the self, the player of the game of chess. The self is defined as a level of consciousness, and its attributes are those of the deep structure of that particular level. The self, Wilber says, is involved in what he calls the Atman-project. Atman (Sanskrit) is the ultimate nature of reality. It is the final unity, the fundamental whole, the source and suchness of reality. It is the top level of consciousness and, at the same time, all levels. It is also indefinable.

The Atman-project is the production of ever-higher unities, the ascension of the ladder of consciousness, encompassing the ladder as the ascent takes place. The goal is to bring all levels into the top level, for a realization that there were no levels in the first place. The self is on all levels all along, but remains ignorant of this by a self-inflicted boundary that limits awareness. The ascent is a widening view of the self that eventually becomes all-encompassing, like the mythic snake swallowing its own tail until nothing remains. That nothingness is everything, and All That Is.

Assuming that the self is engaged in the Atman-project — playing the game of chess, climbing the ladder, or swallowing its tail — how is this done? Does the self lift itself by its bootstraps to the next level? At each level of development, Wilber says, the self maintains itself by a series of constant translations. At each level an appropriate symbolic form emerges and mediates or assists the emergence (through differentiation) of the next higher level. The self then focuses on or identifies with a newly emergent, more complex, and more unified higher structure, which was there all along. According to Wilber's spectrum of consciousness, the biological stage, which is sentient, emerged from the physical or nonsentient material world. Wilber describes the process as follows:

> As evolution proceeds . . . each level in turn is differentiated *from* the self or "peeled off" so to speak. The self, that is, eventually *disidentifies* with its present structure so as to *identify* with the next higher-order emergent structure. More precisely (and this is a very important technical point), we say that the self detaches itself from its *exclusive* identification with that lower structure. It doesn't throw the structure away, it simply no longer exclusively identifies with it. The point is that because the self is differentiated from the lower structure, it *transcends* that structure (without obliterating it in any way), and can thus *operate* on that lower structure using the tools of the newly emergent structure.
>
> Thus, when the bodyego [the mental ego that emerged to create an individual organism] was differentiated from the material environment, it could operate on the environment using the tools of the body itself (such as the muscles).[9]

In effect, Wilber is saying that the self is now on a higher level than the material universe and can use the tools of that higher level to operate in the material universe. The self is now the chess player and can arrange and produce structures through translations of the surface structure. The self is constrained only by the deep structure of the material world and is conditioned by the history of such translations. We should emphasize that although the body is in the material world, and the self can *operate* in and on the material world, it is identified with the next level up. Its activities are now limited by the deep structure of the second level; but when operating within the lowest level, its activities are constrained by the deep structure of that level.

Since it transcends the lower level, the self can both operate in and integrate the lower level. The identification with the next highest level (biological) allows the self for the first time to see that it was identified with the lower level but has now broken away. While it was identified with the material world exclusively, it could not operate in that world since it did not recognize its identification. An example of this process is found in infant development. Wilber observes that at the end of the sensorimotor period of growth, a child can move objects

in a coordinated fashion. However, this is not possible before the child
has become differentiated from its physical environment. As long as
the child is identified with its objective surroundings, it cannot operate on them. As Wilber states:

> At each level of development, one cannot totally see the seer. No
> observing structure can observe itself observing. One uses the struc-
> tures of that level as something . . . to perceive and translate the
> world — but one cannot perceive and translate those structures
> *themselves*, not totally. That can occur only from a higher level.[10]

This concept relates to the problem of bringing an observer (or
consciousness) into physics. Suffice it to say at this point that the self
cannot occupy the level upon which it is acting (see discussion of the
observer in physics, pp. 25-26).

Of the first three levels of consciousness defined by Wilber, the
second, or biological, level is the first one to display the attribute of
life. It is represented by very simple biological systems such as cells.
At this level, there are no concepts, logic, or ideas. The human mind
occupies the third, or mental, level. If the human mind is to operate
on the material level, it must do so by organizing and integrating the
biological level, which, in turn, operates or integrates the material
level. Simply stated, if the mind wishes the body to run or walk, it
must operate and integrate the simpler biological systems of the brain
and legs, which operate and integrate the atoms and molecules of the
material world.

Wilber sees the process of psychological development as mediated
by symbolic forms. These forms assist the self, through differentiation, to rise to the next highest level. A symbol is defined as that
which points to, represents, or is involved with, an element of a different level (either higher or lower). Wilber quotes Huston Smith: "Symbolism is the science of the relationship between different levels of
reality and cannot be precisely understood without reference thereto."[11]
Each deep structure has its own symbolic matrix within which translations of surface structures can unfold and operate. Since the mental
level transcends the material level, the symbols of the mental level
have the power to represent the material level. Language is one of the
symbols used by the child to differentiate itself from the biological
level in order to occupy the mental level. Language provides concepts

that can be used to operate on both the body and the surrounding world. Note that the symbols themselves are *not* material but reside on the mental level. Because of this, they are more creative, more complex, and more unifying; they are not just representations of the material level. The mind is free to operate on or translate these symbols directly without having to perform the inefficient and cumbersome operations on the material level itself.

Wilber's and Seth's views on symbolism differ. Seth sees symbols as reflecting the way we perceive at various levels of consciousness. Seth would agree that each deep structure has its own symbolic matrix, but would insist that the material level in and of itself is a symbol of the mental level. One example might clarify the difference. Seth asks us to consider a fire. To him, a "real" fire is a symbol made physical. We are perceiving a fire with a physically tuned consciousness. On the other hand, a mental picture of a fire is a different symbol. The mental picture may be seen as warmth, but without the destruction, expressing a different feeling. To Seth, symbols express feelings and represent their infinite variations. Physical reality is a symbol of feelings for a level of consciousness that uses physical reality as one of its expressions.

Mathematics is an example of a symbolic system of the mind level. It can be used to represent elements on the material level but, more important, it is more creative, more complex, and more unifying than the material-level elements themselves. We have seen that, according to the Perennial Philosophy, the deep structures of the various levels are one with the ground unconscious, which contains all the deep structures as potentials ready to emerge. This concept may shed some light on the question of why mathematics works in science — and whether mathematics is discovered or created. According to Plato (and more recently Gödel), mathematics is discovered. The Perennial Philosophy would tend to favor such a position since symbols are potentials waiting to emerge. Furthermore, it is not at all surprising that mathematics works in science once the relationship of the levels is considered.

The biological level deals with sensory knowledge, the mind level can be considered the symbolic level, and the subtle level (just above the mind level) is trans-symbolic. Since the mind level is in the symbolic mode, it can form symbols of elements in levels above as well as below. When symbols are used for elements of levels above, the results are eventually paradoxical, since the knowledge of the subtle level is, in essence, direct and cannot be mediated by symbols.

Another way of looking at this is to recognize that symbolic thought is, by its very nature, rational and so requires a subject (knower) and an object (known). Communication on the subtle level being direct, not mediated by symbols or thought, it does not require subject and object. It grasps reality as a whole. Since the symbols of the mental level are inadequate to represent the subtle, the subtle level can never be entirely understood through the concepts, logic, and thought of the mental level. To reach such understanding requires direct access, insight, or transcendence to the subtle level. Even when some insight to the subtle level is obtained through meditation, intuition, or psychic powers, the knowledge still must be described, represented, or symbolized with tools of the mental level in order for the experience to be communicated. This led Huxley to conclude that an operational calculus for the divine Ground is not possible.

The difficulty of symbolizing the subtle level is discussed at some length by Seth. To him, objects themselves are symbols of a physically tuned consciousness. This level of objects is analogous to Wilber's mental level — the stage most humans now occupy. According to Seth, symbols are a method of expressing inner reality when the self is not at higher stages. However, regarding higher stages, he says: "Symbols are no longer necessary, and creativity takes place completely without their use." He describes the ascent to higher levels:

> Beyond this are states in which the symbols themselves begin to fade away, become indistinct, distant. Here you begin to draw into regions of consciousness in which symbols become less and less necessary, and it is a largely unpopulated area indeed. Representations blink off and on, and finally disappear. Consciousness is less and less physically oriented. In this stage of consciousness the soul finds itself alone with its own feelings, stripped of symbolism and representations, and begins to perceive the gigantic reality of its own knowing. . . .[12]

That is, it experiences *directly* — which is beyond the imagination of most of us.

HUMAN DEVELOPMENT:
THE ATMAN-PROJECT AND INVOLUTION

As we pointed out earlier, Huxley stated that the basic premise of the Perennial Philosophy is that the eternal self is one with the Absolute and that each individual is on a journey to discover that fact. Wilber defines this journey as the Atman-project, where Atman is the ultimate reality of nature. Atman is at the top of the ladder of consciousness and at the same time is all the rungs. Each human soul is climbing the ladder to return home, to return to Atman. But since Atman is also the rungs, paradoxically the soul never left home, and needs only to become aware of that fact. Thus, the journey is in actuality the unfolding of higher structures of consciousness. These higher structures are not created, but rather enfolded in lower levels. The question then arises: How did the higher levels become enfolded in the lower ones? This process of enfolding is called involution, as opposed to the Atman-project, or evolution. If involution is the enfolding of higher levels into the lower states, then at the very lowest state (the material level) all of the levels are enfolded as undifferentiated potential.

All this enfolded potential Wilber labels the "ground unconscious." It contains all of the deep structures of all of the levels. At the end of evolution, all of the structures enfolded in the ground unconscious will have unfolded, thus draining the ground unconscious and leaving only Atman. Wilber writes about involution, knowing that the process is beyond description. The following reflects his distillation of a number of mystical traditions[13]:

> The essence of this literature, although it seems almost blasphemy to try to reduce it to a few paragraphs, is that "in the beginning" there is only Consciousness as Such, timeless, spaceless, infinite and eternal. For no reason that can be stated in words, a subtle ripple is generated in this infinite ocean. This ripple could not in itself detract from infinity, for the infinite can embrace any and all entities. But this subtle ripple, awakening to itself, *forgets* the infinite sea of which it is just a gesture. The ripple therefore feels set apart from infinity, isolated, separate.

The creation of the ripple begins the process of involution. At this stage, the ripple is very rarified, or to use Bohm's terminology, extremely subtle. According to Wilber, the ripple is now on the causal level and as such is still quite close to the Absolute. Even so, the first inkling of selfhood is established. It is this sense of selfhood that propels the involutionary process. The self is now paradoxically trapped. On the one hand, it wishes to return to the Absolute to restore its profound peace. But to do so, it literally must die — it must give up its sense of self, a terrifying prospect. As a result, the self seeks fulfillment by a compromise. As Wilber puts it:

> Instead of finding actual Godhead, the ripple pretends itself to be god, cosmocentric, heroic, all-sufficient, immortal. This is not only the beginning of narcissism and the battle of life against death, it is a *reduced* or *restricted* version of consciousness, because no longer is the ripple *one* with the ocean, it is trying itself to be the ocean.

The Atman-project then becomes the reverse — the involution process. The ripple creates more restricted orders of consciousness. That is, it descends from the causal level of perfect transcendence to the subtle level by reducing the scope of its consciousness. But its desire for infinity is not satisfied at this level. So its scope is again reduced — to the mental level. The process is repeated until the ripple falls to the material level and goes into "insentient slumber." At this point, Wilber inserts the following reminder:

> Yet behind this Atman-project, the ignorant drama of the separate self, there nonetheless lies Atman. All of the tragic drama of the self's desire and mortality was just the play of the Divine, a cosmic sport, a gesture of Self-forgetting so that the shock of Self-realization would be the more delightful. The ripple *did* forget the Self, to be sure — but it was a ripple *of* the Self, and remained so throughout the play.

This entire involution process enfolds all the structures in the ground unconscious. The stage is now set for the reverse process, the Atman-project. The deep structures now contain all the undifferentiated potential needed to return to Atman. Instead of restricting consciousness, the self must recognize that it is actually one with the Absolute and thereby expand its consciousness through its journey.

Wilber concludes that the evolution of the universe can be compared with the Atman-project. At the time of the big bang, all was on the material level. That was followed by matter broadening its consciousness to primitive life forms on the biological level. Then, by a further awakening, the process achieved level three, or human existence. Wilber quotes Plotinus: "Mankind is poised midway between the gods and the beasts." Thus the process of involution created the conditions for evolution and the Atman-project.

MICROGENY: INVOLUTION MOMENT BY MOMENT

The involutionary process not only occurred at the beginning but, according to the Perennial Philosophy, continues to occur in two other ways. The first takes place at biological death; it is described in *The Tibetan Book of the Dead*. The second happens moment by moment and relates directly to our thesis.

Many spiritual traditions describe the journey of the soul from the moment of death to the moment of rebirth. *The Tibetan Book of the Dead* calls it the Bardo state and says that it can take up to 49 days. Although the time element appears to be contrary to the moment-by-moment involutionary process, the Tibetan tradition considers the two processes to be the same. At physical death, the soul enters into the top level of consciousness, or Atman. (That is, the soul was there all the time but now becomes aware of it.) Then begins the process downward from the top. Because of its "karmic propensities," the soul contracts away from Atman because it cannot stand the intensity of pure Oneness. This process of contraction is repeated through the various levels of consciousness. The transformations are downward until rebirth occurs in a new body. Then, in a new body, the soul begins its journey upward again, hoping to rid itself of its karmic propensities so that at its next death and rebirth, the fall will end at a higher level.

Karma is seen as the collective force of an individual's past action, a summary of causal factors in the area of moral conduct. It determines the level at which the involution, or fall, ends. During this process, the soul is accompanied by a swoon of forgetfulness, and the entire sequence becomes unconscious. In other words, the higher levels are all there, but they must be remembered and unfolded from the ground unconscious or the deep structures of each level.

The involutionary process also takes place moment by moment. Wilber describes it as follows:

> Here, finally, is the other meaning of the Bardo, of the In Between, and if you feel that "reincarnation" or "rebirth" is unacceptable, then this might be easier to accept (although they both are really *exactly* the same): not only did the whole involutionary series occur prior to one's birth, one re-enacts the entire series moment to moment. In this moment and this moment and this, an individual *is* Buddha, *is* Atman, *is* the Dharmakaya — *but*, in this moment and this moment and this, he ends up as John Doe, as a separate self, as an isolated body apparently bounded by other isolated bodies. At the beginning of *this* and every moment, each individual *is* God and the Clear Light; but by the *end* of this same moment — in a flash, in the twinkling of an eye — he winds up as an isolated ego. And what happens In Between the beginning and ending of *this* moment is identical to what happened In Between death and rebirth as described [in the *Tibetan Book of the Dead*].[14]

Wilber calls this process "microgeny." It is a moment-by-moment reenactment of the Bardo state as defined in *The Tibetan Book of the Dead* — but with a vastly different time element. While the Bardo state can last 49 days, the time of the microgeny process is incredibly short. As we shall see in Chapter 5, it may involve time spans as short as 10^{-43} seconds. But as in the Bardo sequence, the self passes through all the levels above it, and on the return trip ends up at the level to which it has evolved. As an example, if the self is on the mental level, it passes through the subtle and causal levels and on to the ultimate. On the return, the self settles again on the mental level. But each moment — for a split second — each individual is the Buddha. However, the self does not remember this exposure to the higher levels. The remembrance can only occur through the Atman-project or evolution of the individual. Again, it must be emphasized that the self does not go anywhere in space; there is an expansion and then a contraction of consciousness. All levels are right here, right now.

In essence, we have the whole process of evolution and involution occurring not only over long stretches of time but also at each instant. The whole array of higher levels actually generates the lower levels every moment through an interpenetration of all levels with each other.

This means that the whole spectrum of consciousness was not just created during the initial involution process but is actually recreated each split second. If one were to select the material level as an example, this suggests that our material universe was not only created by a big bang many years ago, but there is also a big bang happening moment by moment. It should be noted that when the universe is "not here" for a fraction of a second, it is part of the Absolute — or part of the deep structures of all the levels. Rather than give an exact time span for the cycle of this process, Wilber uses the terms "in a flash" or "in the twinkling of an eye." In any case, this phenomenon of microgeny is a projection of consciousness into a spectrum and an injection back into the Absolute at an extremely high frequency. The way microgeny relates to contemporary physics and Seth will be explored in Chapter 5.

TIME AS A PRODUCT OF THE MENTAL LEVEL

According to the Perennial Philosophy, the concept of time (and space) is a product of the mental level in the spectrum of consciousness. Wilber points out that the first two levels, the physical and biological, are timeless, but in a pretemporal sense. That is, the concept of time has not yet been grasped because the levels are too primitive. In contrast, the levels above the mental have transcended time because the need for such a construct has diminished or disappeared. On the level of the physical (the material world), the concept of time is completely nonexistent. On the biological level (prelanguage), the concept of time is what Wilber calls prelinear. It is on the mental level, with its symbolic structure of language, that the structure of past and future, called time, is created.

Seth describes the relationship of time and language in this way:

> In the first place, language as you know it is a slow affair: letter by letter strung out to make a word, and words to make a sentence, the result of a linear thought pattern. Language, as you know it, is partially and grammatically the end product of your physical time sequences. You can only focus upon so many things at one time, and your language structure is not given to the communication of intricate, simultaneous experience.[15]

The highest level of the spectrum of consciousness is timeless and spaceless. This concept is difficult to grasp since we are products of the mind level. But as we have seen from Bohm's concept of the hologram, the infinite is present at every point of space. In Wilber's discussion of this[16] he says:

> Since the infinite is present in its entirety at every point of space, *all* of the infinite is fully present right HERE. In fact, to the eye of the infinite, no such place as *there* exists (since, put crudely, if you go to some other place over *there*, you will still only find the very same infinite as *here*, for there isn't a different one at each place).

Similarly, the Absolute is present at every point in time since the Absolute is timeless.

> Being timeless, *all* of Eternity is wholly and completely present at every point of time — and thus, all of Eternity is already present right NOW. To the eye of Eternity, there is no *then*, either past or future.

Thus, eternity bears the same relationship to time as infinity does to space: all of time is at the present moment, and all of space is at each point in space. This means that each point in space is identical to all other points, since all points contain the infinite. In these terms, space is nonexistent. In a similar manner, only the time "now" exists. Past and future are human constructions.

Another way of looking at this difficult concept is to envision the entire Absolute being present at every point of space and time. The following example may help. In our three-dimensional universe, an event may occur at point A. If the knowledge of that event is transmitted with the speed of light, it arrives at point B some time later. This information can be transmitted only within a time *not less* than it takes for light to travel from A to B. This is the limiting velocity in our three-dimensional world. However, if we operated on a level where this velocity restriction were not present, then point B could receive the information from point A instantaneously. In such a situation, every point is connected to every other point without a time restriction. The corollary is also true: all points in space are contained in each point. As Wilber puts it, "The *entire* Absolute is completely and wholly present at every point of space and time . . . ."[17]

The concept of time derives from the fact that we divide up reality into bits and pieces. In so doing, we create a subject and an object with a space in between. We experience these objects, these aspects of reality, in a piecemeal manner and thus create a linear succession in time. The spaceless infinite removes the space between subject and object; the timeless now removes linear succession. Instead of knowing the universe from a distance, the self on the higher levels knows the universe by *being* it — without need of space and time.

SUMMARY

Wilber states that a hierarchy of consciousness is the main thrust of the Perennial Philosophy; all the mystical traditions of the major religions exhibit some sort of hierarchy of levels. The higher levels enfold or include the lower ones. This hierarchy should not be viewed as a linear or sequential ladder. The entire range of levels is in fact not hierarchic at all; it is infinite and not qualifiable. The spectrum is merely a manifestation of the whole, which is all-inclusive; indeed, the levels can be viewed as illusory. All levels are mutually interpenetrating and interconnecting; all levels are conscious, the lowest level being called insentient consciousness.

A further concept is that each level can be divided into a deep structure and a surface structure. The deep structure contains all the potentials and limitations of that level. The surface structure is a particular manifestation of the deep structure.

Wilber emphasizes that the levels are *not* separate spatial regions, but interpenetrate one another, with the higher levels forming and sustaining the lower levels. He defines the ground unconscious as all the deep structures with all potentials ready to manifest in the surface structure. Wilber further points out that the self identifies with the symbolic structure emerging out of a given level and transcends that level. Only in this manner can it operate on the lower level and integrate it.

The concepts of hierarchy, enfoldment of levels, and the infinite nature of the hierarchy all are similar to concepts developed by Bohm. The notion of the deep and surface structures obviously parallels Bohm's implicate and explicate orders. The ground unconscious and the holomovement have common elements. Also, Wilber's concept of microgeny and Bohm's unfolding and enfolding are comparable. Bohm, using the concepts developed in contemporary physics, and Wilber, using an amalgamation of mystical traditions, have arrived at amazingly similar ideas. When Seth is brought into the comparison, the analogous aspects of these views are even more striking.

3

Seth:

A Paranormal

Perspective

Reality may not only be stranger than we conceive, it may be stranger than we can conceive.

J. B. S. Haldane

Seth, the discarnate entity channeled by Jane Roberts, is the third of our sources. His view of reality is not consonant with our everyday sense of how things are. To better understand his concepts, we must widen our focus and attempt to see reality through his lens.

Seth in turn makes every effort to speak to us with our terminology.[1] Like the fictitious entity Gezumpstein in Alex Comfort's *Reality and Empathy*, Seth does not observe events in a time dimension, but sees all happenings simultaneously en bloc. To communicate with us, Seth must continually translate his thoughts into our sequential mode of thinking. Translation, of course, can lead to distortions. In addition, as Seth readily admits, he does not have all the answers. As we have learned from David Bohm, no level of order has complete information.

In *The Seth Material*, Jane Roberts presented a discussion between Seth and Eugene Barnard, an academic psychologist with extensive knowledge of Eastern philosophy. Dr. Barnard said afterward:

I chose topics of conversation which were clearly of tolerable interest to Seth and considerable interest to me, and which by that time I had every reason to believe were largely foreign territory to Jane. Also . . . I chose to pursue these topics at a level of sophistication which I felt . . . made it exceedingly improbable that Jane could fool me on [by] substituting her own knowledge and mental footwork for those of Seth, even if she were doing it unconsciously. . . .

The best summary description I can give you of that evening is that it was for me a delightful conversation with a personality or intelligence or what have you, whose wit, intellect, and reservoir of knowledge far exceeded my own. . . . In any sense in which a psychologist of the Western scientific tradition would understand the phrase, I do not believe that Jane Roberts and Seth are the same person, or the same personality, or different facets of the same personality. . . .[2]

Whether or not one accepts Barnard's views, it will be advantageous to our discussion to consider Seth as a separate entity from Jane Roberts. Seth brings a unique perspective to our search for a new paradigm. His views will be compared with those of Bohm, Wilber, and others representing a variety of disciplines.

THE EGO, A MULTIDIMENSIONAL PROJECTION

Carl Jung's prodigious study of dreams, fantasies, myths, and legends revealed a striking similarity between the images and stories of peoples widely separated geographically and historically. Jung concluded that these images spring from the same forms residing in an underlying collective level of awareness. He called this level the collective unconscious. To Jung, the collective unconscious, as opposed to our personal everyday consciousness, is a second psychic system available to all individuals. It is universal, impersonal, and largely inherited. It contains pre-existent forms, or archetypes, which become conscious only in a secondary manner. The archetypes can be thought of as patterns of instinctual behavior.

Although Seth views consciousness as the basis of all reality, he has also used the terms "unconscious mind" and "collective unconscious." He agrees with Jung that the outer or "normal" ego is not aware of unconscious material directly. In Seth's language, outer ego refers to our normal waking consciousness, which operates in the three-dimensional world. The inner ego (another term for the inner self) organizes the inner reality of the collective unconscious, thereby creating the substance that we call the three-dimensional world. The inner ego is a higher level of consciousness, much more active and knowledgeable than the outer ego. The outer ego deals day by day with the physical reality created by the inner ego.

The inner ego (or unconscious mind) is both purposeful and highly discriminating. Seth says that the outer ego rises out of the inner ego; the outer self is the offspring of the inner self. The inner ego has access to a huge library of knowledge (the collective unconscious), and through its creation of the physical world provides stimuli to keep the outer ego constantly alert and aware. The inner ego projects a material reality in which the outer ego can act out its role. Seth uses the analogy of the outer ego acting out a play that the inner ego has written (pointing out, however, that the outer ego should not be perceived as a mere puppet).

In a sense, the outer ego is spoonfed — given only that information, including feelings and emotions, that it can handle. This information is *usually* in the form of data picked up by the physical senses. Matter, in short, is the shape that basic experience takes when the inner ego projects into the three-dimensional world. The similarity between the collective unconscious and the implicate order is apparent. Jungian analyst Marie-Louise von Franz has this to say:

> For the psychologist it is clear that in his idea of an "implicate order" David Bohm has outlined a projected model of the collective unconscious, so that in his theory we have before us an attempt to outline a psychophysical model of the unity of all existence. The background of this existence, as Bohm expresses it, is an infinite reservoir — a "vast sea" — of energy which lies deeply behind/under our consciousness, which is unfolded in space-time.[3]

Seth frequently points out that the outer ego's restrictions are self-imposed, resulting from its own ignorance and limited outlook. The conscious mind is not some prodigal child or poor relative of the inner self. It can see into inner reality once it understands that it is possible to do so and rids itself of unnecessary limitations.

The outer and inner egos together organize and manipulate an immense repository of knowledge to produce the three-dimensional world. Here is more definite resemblance to Bohm's ideas. Bohm's superimplicate order organizes the vast treasury of knowledge of the implicate order also to produce the three-dimensional world. The superimplicate order guides a particle through its three-dimensional trajectories just as the outer and inner egos, in combination, guide the human body through its movements. In a similar way, the self of the Perennial Philosophy organizes the underlying deep structure to give form to the surface structure. Whether we are discussing a particle, a cell, an animal, or a human being, a framework corresponding to explicate, implicate, and superimplicate orders makes its appearance.

An alternative way of viewing the concept of the inner and outer egos is found in Bohm's notion of an infinite spectrum of orders. The outer ego has available to it a more restricted information system, presumably to enable it to accomplish its task more effectively. The inner ego has a more extensive realm of information and therefore a broader context. An anomaly may not be explainable to the outer ego within its system of information. But with a wider focus, the anomaly can be understood. All this is not to denigrate the outer ego. Within its range, its information is correct, though limited. As Seth continually assures us, a broader focus is available to the outer ego when and if the ego learns to handle it, that is, through the development of certain psychic faculties.

In some sense, Bohm's theory contains an analogue to the inner and outer egos. Recall that the implicate order is viewed as multidimensional and the explicate order as three-dimensional. Mathematically, the distinction can be made by terming the implicate order $3n$-dimensional and the explicate order three-dimensional. The term n can be any whole number from one to infinity. Thus, the unmanifest portion of reality is a spectrum of orders defined by a given value for n.

Bohm sees matter as a manifestation in three dimensions from a multidimensional source. (Recall his analogy of the fish in a glass tank filmed by two video cameras on different sides of the tank. When the films are projected onto two video monitors, it appears that there are two fish, moving in different directions. That is, when the three-dimensional reality is projected onto two dimensions, an illusion results in which a single entity is perceived as two separate entities.) Essentially, the laws of quantum mechanics relate the multidimensional order to our observed three-dimensional order. In a similar manner, Bohm says there is the equivalent of a $3n$-dimensional consciousness and a three-dimensional consciousness. The three-dimensional consciousness is certainly more subtle than matter but is limited compared with the $3n$-dimensional consciousness. Our everyday consciousness (three dimensions) deals in "thought" as we normally define it. The $3n$-dimensional consciousness is more aware, more intelligent, and certainly more intuitive. Meditation, Bohm feels, utilizes this higher consciousness. The everyday consciousness needs to expand its view beyond the explicate order — beyond thought — so it can break free from the restraints of a three-dimensional reality. There are obvious parallels between three-dimensional/$3n$-dimensional consciousness and outer/inner egos.

Bohm's concept of a $3n$-dimensional consciousness is a useful analogy in understanding one of Seth's most frequently repeated views, that "the soul is not a closed system." We noted earlier that the outer ego springs from the inner ego. The inner ego is a higher-dimensional consciousness, which cannot be contained in three-dimensional reality. The inner ego has a higher value for n than the outer ego (for which the value of n is obviously 1). However, that is not the end of the story: the system is not closed. The number of dimensions, the value of n, can reach infinity. That implies that the inner ego is itself a projection of a still higher consciousness. This higher consciousness Seth calls "the soul" or "entity." To quote:

> The "outer ego" and the inner ego operate together, the one to enable you to manipulate in the world that you know, the other to bring you those delicate inner perceptions without which physical existence could not be maintained.

> There is however a portion of you, the deeper identity who forms both the inner ego and the outer ego, who decided that you would be a physical being in this place and in this time. *This* is the core of your identity, the psychic seed from which you sprang, the multidimensional personality of which you are part.[4]

The soul, of course, has an even higher value for n than does the inner ego. Again we are describing a hierarchy of consciousness that seems to imply divisions between the various levels, which Seth emphatically denies:

> You must understand that there are no real divisions to the self . . .
> so we speak of various portions only to m make the basic idea clear.[5]

All consciousness gestalts are projections of a higher gestalt, but basically they are *one* consciousness. As n is increased, the consciousness at a given level is simply not containable in a lower-dimensional reality. The soul can more fully express itself and grow by using lower values of n, and an infinite n is equated with All That Is. We will discuss this concept more fully in the section on the spectrum of consciousness in this chapter.

The notion of outer and inner egos has some intriguing correspondence to the results of experiments performed by Ernest R. Hilgard. Hilgard, an experimental psychologist, and his wife Josephine Hilgard, a psychiatrist, are distinguished researchers in the field of hypnosis. In their book *Hypnosis In The Relief Of Pain*, they propose that in a metaphorical sense there is a "hidden observer" that can be distinguished from the conscious mind.[6] They describe how during a classroom demonstration, a blind student was hypnotized and told that at the count of three he would be totally deaf to all sounds. Of course, being blind, the subject could not make use of any visual cues. While he was under hypnosis, wooden blocks were banged near his ears and a starter's pistol was fired near him. He showed no response. Then Hilgard addressed the student as follows:

> As you know, there are parts of our nervous system that carry on activities that occur out of awareness, of which control of the circulation of the blood, or the digestive processes, are the most familiar. However, there may be intellectual processes also of which we are unaware, such as those that find expression in night dreams. Although you are hypnotically deaf, perhaps there is some part of you that is hearing my voice and processing the information. If there is, I should like the index finger of your right hand to rise as a sign that this is the case.

To everyone's surprise, the finger rose. Then the student said:

Please restore my hearing so you can tell me what you did. I felt my finger rise in a way that was not a spontaneous twitch, so you must have done something to make it rise, and I want to know what you did.

Still under hypnosis but with his hearing restored, the student was asked what he remembered about the incident. He answered:

I remember your telling me that I would be deaf at the count of three, and could have my hearing restored when you placed your hand on my shoulder. Then everything was quiet for a while. It was a little boring just sitting here, so I busied myself with a statistical problem I have been working on. I was still doing that when I suddenly felt my finger lift; that is what I want you to explain to me.

The student was then brought out of hypnosis.

The account of the hypnotized student indicates that at some level, a portion of his consciousness was able to understand Hilgard while the "normal" consciousness appeared to be totally deaf. In later experiments dealing with covert and overt pain, this "hidden observer" was again encountered.

Hilgard also discusses a study involving age regression and hypnosis. Adult subjects were asked to go back to childhood and experience being five years old. Hilgard reports:

This experience takes two different forms. In one form the subject becomes completely absorbed in the experience of being a child again, while in the second form the subject becomes a child again in a manner that feels convincing, but in addition there is an observer present. This observer has some of the properties of the hidden observer in pain in that it knows all that is going on in the inward experience as well as in the environmental contexts of the experience.[7]

Jeremy Campbell comments on other experiments with hidden observers:

The hidden observer proved to be superior, in many ways, to the conscious individual inside whom it was hidden; more mature, more realistic and logical, and *in possession of more information*. One student remarked that he could let his imagination run free in a hypnotic trance and indulge in some pleasing fictions, "but somewhere the hidden observer knows what's really going on."[8]

The outer and inner egos have counterparts in Buddhism also. Note this statement by the Dalai Lama:

> Buddhism teaches that we have six senses: the consciousness of sight, hearing, smell, taste, touch, and mental consciousness or "mind-consciousness"; the latter refers to the ordinary, daily mind that registers, relates, combines, and stores sense-impressions. This is not the highest stage of mind or consciousness.[9]

Mind-consciousness is similar to the outer ego and performs the same functions. The statement that there is a higher stage of consciousness implies some counterpart to the inner ego.

Seth frequently states that the physical objects we perceive are actually symbols, created by us in a fashion similar to the way in which we create words. Just as a printed page does not *contain* information but *transmits* information, so do physical objects represent reality but are not reality themselves. Objects are natural by-products of the evolution of our species, just as language is. In the same way that words on a page are translated into a counterpart of thought, objects are created as a method of communication. What we perceive, then, is a form of language we call physical reality — a language we learned as infants and use automatically. Physicist Roger S. Jones arrived at this very understanding via a different route:

> I had come to suspect, and now felt compelled to acknowledge, that science and the physical world were products of human imagining — that we were not the cool observers of that world, but its passionate creators. We were all poets and the world was our metaphor.[10]

The objects we observe have solidity only in our three-dimensional environment. Anyone existing outside this environment does not perceive such objects, and conversely, such objects presumably can exist unrecognized alongside us.

Most researchers in the field of developmental psychology agree that the newborn infant does not separate subject and object; that is, it does not have a sense of self as separate. Wilber points out that it is not that the infant is born into a world of objects that cannot be recognized; rather, from its perspective, there are no objects, because there is no separation from the totality of its experience. Seth would say that the ego of the infant has not yet mastered the skill of creating objects; it has not yet learned to express thoughts as material things.

Why is it necessary that the outer ego perform in the world at the direction of the inner ego? The answer, according to Seth, is that we are learning to translate energy and ideas into experience.

> In your system of reality you are learning what mental energy is, and how to use it. You do this by constantly transforming your thoughts and emotions into physical form. You are supposed to get a clear picture of your inner development by perceiving the exterior environment. What seems to be a perception, an objective concrete event independent from you, is instead the materialization of your own inner emotions, energy, and mental environment.[11]

In summary, objects that we perceive are actually created by us (by the inner ego) and are symbols that express our emotions and thoughts. It is through this process that we develop our consciousness.

BASIC BUILDING BLOCKS: ELEMENTS OF CONSCIOUSNESS

Seth points out that thoughts and emotions are formed into physical matter by definite methods and through specific laws not presently known to us. Throughout the books of Jane Roberts, he fills in the details of this remarkable process.

In order to help us understand matter, Seth describes basic building blocks that he calls "units of consciousness," or CUs. To say the least, these units are extraordinary. Since they are not physical and do not occupy space, they can be in all places at all times simultaneously. This freedom from three-dimensional constraints permits them to be precognitive and clairvoyant. The CUs contain within themselves innately infinite properties of expansion, development, and organization, yet they always maintain the kernel of their own individuality. They *do not* have human characteristics, but they do have their own leanings, inclinations, and propensities. They are not idle energy, but are vitalized, aware, charged with all the qualities of being. The basic CU is endowed with unpredictability, which allows for infinite patterns and fulfillments. In a sense, the CUs are divine fragments of All That Is.

That consciousness is the basic constituent of the universe — that at its very essence the world is in some sense alive — is expressed (in different words) by Seth, Bohm, and Schrödinger. In reference to CUs, Seth says, "Despite whatever organizations it becomes part of, or how it mixes with other such basic units, its own identity is not annihilated."[12]

Bohm makes this statement:

> As you probe more deeply into matter, it appears to have more and more subtle properties. . . . In my view, the implications of physics seem to be that nature is so subtle that it could be almost alive or intelligent.[13]

In *My View of the World*, Schrödinger says:

> Consciousness is that by which this world first becomes manifest, by which indeed, we can quite calmly say, it first becomes present; that the world *consists* of the elements of consciousness. . . .[14]

Seth makes it clear that CUs are not to be thought of strictly as particles, but that they can operate as either particles or as waves. To use Bohm's terminology, the CUs act as particles when they are explicated; when they are in the implicate order, they act as waves. This wave-particle description of a CU allows us to utilize a metaphor for consciousness that does not restrict us to a particulate description. That is, consciousness need not be sharply localized within a given body. If consciousness has a wave as well as a particle nature, then certain psychic phenomena begin to make sense. As an example, such clairvoyant effects as remote viewing might be explained as a reflection of the wave aspect of consciousness.

When Seth speaks of CUs as individual units, he is referring to only a part of the total psychic energy that he calls All That Is, a term similar to Bohm's holomovement. The following passage by Seth helps clarify this point:

> You must realize that pure energy has such transforming pattern-forming propensities that it always appears as its manifestations. It becomes its "camouflages."

> It may form particles, but it would be itself whether or not particles existed. In the most basic of terms, almost incomprehensible in your vocabulary, *energy is not divided*. There can be no portions or parts

of it, because it is not an entity like a pie, to be cut or divided. For purposes of discussion, however, we must say that in your terms each smallest portion — each smallest unit of pure energy — contains within it the propelling force toward the formation of all possible variations of itself.

The smallest unit of pure energy, therefore, weighing nothing in your terms, containing within itself no mass, would hold within its own nature the propensity for the creation of matter in *all* of its forms, the impetus to create *all* possible universes. In those terms, energy cannot be considered without bringing to the forefront questions concerning the nature of God and All That Is, for the terms are synonymous.[15]

The wholeness that Bohm describes in the holomovement is echoed in Seth's comment that energy cannot be divided. We only appear to do so by viewing certain aspects while ignoring others. CUs can take on form and become separate entities, or they can flow together in a vast, harmonious wave of activity as a force. Actually, Seth says, they operate in both ways at once:

Each "particleized" unit . . . rides the continual thrust set up by fields of consciousness, in which wave and particle both belong. Each particleized unit of consciousness contains within it inherently the knowledge of all other such particles — for at other levels, again, the units are operating as waves.[16]

There are two elements in the above excerpt that are analogous to Bohm's causal interpretation of quantum theory. Note that when the CU acts like a particle, it rides the wave of fields of consciousness to which wave and particle both belong. In this sense, the concept is similar to the wave function guiding the particle through the quantum potential. In another sense, the wave and the particle are one, the particle being just a localized manifestation of the quantum field. Also, because of the quantum potential, the particle, Seth says, "contains within it inherently the knowledge of all other such particles." This reminds us of a hologram.

In effect, Seth is saying that for purposes of analogy, the CUs can be considered the building blocks of reality. This concept has an interesting parallel in the history of physics. In the nineteenth century — the heyday of classical physics — the world was thought to be made up of two substances or entities, matter and fields. Both had reality in the three-dimensional world, and the relationship between the two was well defined. Matter produced fields of force (in contradistinction to information fields or even probability fields). The force was distributed throughout space and fell off in intensity with distance. The fields were identified as gravitational and electromagnetic. In the twentieth century, other fields were added that did not spread throughout space, but the fundamental assumptions of classical physics did not change. Reality consisted of matter and fields, and basic laws were known for both.

Quantum theory essentially erased the difference between matter and fields, making reality a unity that exhibits the properties of both. This single, unitary stuff gave rise to the fantastically successful algorithm now used by physicists in all calculations involving quantum theory. But nobody knows what this unitary stuff really is. Seth, of course, defines it as CUs, which have essentially the same attributes as the quantumstuff (wave and particle attributes). Most quantum physicists, of course, stop short of calling this unitary substance consciousness.

Seth explains that CUs can move faster than light and slow down to the speed of light to form matter. Before matter appears, there is a "disturbance" in the spot of space-time where the materialization is to take place. The slowing down of prior faster-than-light activity helps "freeze" the activity into a form. The initial faster-than-light activity and the deceleration cannot be ascertained from within the three-dimensional universe. In this connection, Bohm says:

> Mass is a phenomenon of connecting light rays which go back and forth, sort of freezing them into a pattern.

> So matter, as it were, is condensed or frozen light. Light is not merely electromagnetic waves but in a sense other kinds of waves that go at that speed. Therefore all matter is a condensation of light into patterns moving back and forth at average speeds which are less than the speed of light. Even Einstein had some hint of that idea. You could say that when we come to light we are coming to the fundamental activity in which existence has its ground, or at least coming close to it.[17]

Characteristics common to Bohm's frozen light and Seth's CUs are explored in detail in Chapter 5.

COORDINATE POINTS AS
CONSCIOUSNESS TRANSFORMERS

Objects have solidity only within our three-dimensional world. Other universes may be considered to exist alongside us, unperceived, outside our reality. There are points, however, at which different realities do coincide. In Seth's words,

> These coordinate points . . . act as channels through which energy flows, and as warps or invisible paths from one reality to another. They also act as transformers, and provide much of the generating energy that makes creation continuous in your terms. . . .[18]

The closest analogy to coordinate points in physics today is the concept of the black hole and white hole. While the relationship between coordinate points and the black hole/white hole will be discussed later (Chapter 5), we might point out here that a black hole is a hole in space with a definite edge where matter can literally fall out of our universe. The white hole is its exact counterpart, a hole in space where matter escapes into our universe.

According to Seth, all coordinate points represent accumulations of pure energy. This energy is dormant until activated, and it cannot be activated physically. Seth claims that there is an "ever-so-minute" alteration of gravity and a wavering of the effect of physical laws in the neighborhood of any of these points.

Seth describes three types of coordinate points: absolute coordinate points that intersect *all* realities, main coordinate points, and subordinate points. The four absolute coordinate points are sources of fantastic energy, but the subordinate points are far more common and affect our daily concerns more directly. These concentrated energy points are activated by our thoughts (and emotions), and the energy of the original thought is reciprocally enhanced by the coordinate point. The intersections of thought and coordinate point make possible the projection of the thought into physical matter. The nature of the thought is not important, only its intensity, Seth says:

These points are like invisible power plants, in other words, activated when any emotional feeling or thought of sufficient intensity comes into contact. The points themselves intensify whatever activates them in a quite neutral manner.[19]

Bob Toben and Fred Alan Wolf come to a similar conclusion:

I think that thought does modify the strength of quantum wave functions. Now, the strength of a quantum wave is a measure of the probability of an event occurring. I believe that the greater the awareness or consciousness of the observer, the greater the probability of the event occurring.[20]

The subordinate points also serve as supports within the unseen fabric of energy that forms all realities and manifestations. This suggests an infinite energy pool in which all realities exist. Seth defines the energy as a divine gestalt of being of unimaginable dimensions, a vast field of multidimensional creativity. Bohm offers a similar description: "There is an implicate order and there is an infinite sea of energy and . . . this unfolds to form space, time and matter."[21]

Subordinate points *might* be related to the dark matter of modern cosmology. Simply stated, physicists infer the existence of hidden matter spread throughout the universe. It is considered dark because it largely does not interact with our familiar matter and therefore is not detectable by our instruments. But it does have gravitational effects, used to explain the clustering of galaxies (among other things). Some estimates place the amount of dark matter at 90% of the total matter in the universe; only 10% has been detected. When we consider Seth's description of subordinate points — extremely small points that contain pure dormant energy and cause minute alterations of gravity nearby — we cannot help but wonder how this might relate to the concept of dark matter.

MATTER FORMATION: SETH'S METAPHOR

Seth's explanation of matter formation — which, he cautions us, is highly simplified — is that solid matter is formed through the relationship of CUs and coordinate points. A combination of CUs forms a consciousness. The subjective experience of consciousness automatically expresses itself as electromagnetic energy (EE) units. The subjective experience that creates them can be a thought, a mental image, an emotion, a stimulus, or a reaction (we will simply say "mental activity"). A strong emotional component makes it more likely that mental activity will be manifest in physical form. The EE units are natural emanations from consciousness. They exist not in three-dimensional space but just beneath the range of physical matter. They seldom exist in isolation but unite under certain conditions. They pulsate and can change their form. Their duration depends on the intensity of the mental activity from which they originate; if the intensity is high enough, the EE units activate the dormant energy of a subordinate/coordinate point and are propelled into matter. A similar idea was expressed by the English philosopher and mathematician Alfred North Whitehead: "The energetic activity considered in physics is the emotional intensity entertained in life."[22]

The EE units emitted by consciousness can be compared to Bohm's concept of meaning, which he expresses as follows: "Meaning is something like the form which informs the energy, so it will actively direct the energy and shape it."[23] It is important to note that consciousness is shaping consciousness, that matter is a form of consciousness. As Bohm also says:

Mind and matter are inseparable, in the sense that everything is permeated with meaning. The whole idea of the somasignificant or signasomatic is that at no stage are mind and matter ever separated. There are different levels of mind. Even the electron is informed with a certain level of mind.[24]

When EE units are operating beneath the surface of reality, they are traveling at a velocity greater than light and therefore do not emerge as matter in our three-dimensional universe. When they do become matter — through the help of coordinate points — they are slowed to the velocity of our light and then take on solid form. A more detailed description of this process is found in Chapter 5.

In short, EE units originate as mental activity that, under certain conditions and with the help of coordinate points, emerges as the physical building blocks of matter. Mental activity provides the blueprint for the object, condition, or event in three-dimensional space. In a manner similar to the way in which the sun aids the growth of plants, coordinate points and EE units, with consciousness as a guide, activate the behavior of atoms and molecules. They encourage the tendency of units of matter to "swarm" into various forms or structural groupings. The EE units are not confined to human consciousness but arise from all levels of consciousness, including the cellular.

The EE units pervade our entire space beneath the surface of our reality, like an ocean of invisible formations, analogous to the vast energy pool. Under certain conditions, determined by all consciousness, islands of physical matter are formed and guided by the invisible formations. Using Bohm's terminology, the EE units can be compared to "active" information directing a much greater energy source, the subordinate point. Bohm explains:

> The basic idea of active information is that a *form*, having very little energy, enters into and directs a much greater energy. This notion of an original energy form acting to "inform," or put form into, a much larger energy has significant applications in many areas beyond quantum theory.[25]

Bohm uses the example of a radio signal. The form of the radio wave carries information such as the voice of the announcer. But the sound we hear does not receive its energy from the signal; the energy comes from the power socket. The energy is initially unformed but develops form from the information fed into the receiver from the antenna. The information from the station is then *potentially* active everywhere the signal can reach. It becomes *actually* active upon being received by the radio antenna. Bohm says:

> The analogy with the causal interpretation is clear. The quantum wave carries "information" and is therefore *potentially* active everywhere, but it is *actually* active only when and where this energy enters into the energy of the particle. But this implies that an electron, or any other elementary particle, has a complex and subtle inner structure that is at least comparable with that of a radio. Clearly this notion goes against the whole tradition of modern physics, which assumes that as matter is analyzed into smaller and smaller parts, its behavior grows more elementary. By contrast, the causal interpretation suggests that nature may be far more subtle and strange than was previously thought.[26]

The subordinate points and EE units, as they intrude into our three-dimensional universe, make up the vast energy pool of space.

This space is essentially passive; it cannot create action in and of itself. In Bohm's terminology, it is linear. It requires a form that is more subtle and more complex, like the superquantum potential, to modify its field so that it can become nonlinear. Mental activity, which is required to activate the subordinate points, is analogous to the superimplicate order in that it provides the stimuli to form the field into matter. Bohm observes: "There is a similarity between thought and matter. All matter, including ourselves, is determined by 'information.' 'Information' is what determines space and time."[27]

As stated before, CUs move faster than light but slow down to form matter. Mental activity, as EE units, forms matter with the energy of the subordinate/coordinate points. All activity stays below the surface of reality except for the resultant material objects. Mystic philosopher P. D. Ouspensky sums this up very well: "The true *motion* which lies at the basis of everything is the *motion of thought*. True *energy* is the *energy of consciousness*."[28]

MATTER FORMATION: A PHYSICS METAPHOR

Let us briefly explore another conception of reality, which involves a wave function traveling backward and forward in time. Developed by physicist John Cramer, the Transactional Interpretation has some similarities to Seth's ideas, although, of course, its terminology is derived from quantum theory. Wolf uses the Transactional Interpretation in his book *Star Wave*, with a twist that brings it more in line with Seth's description.

To begin, let us define the event we are attempting to describe as a system going from energy state A to energy state B. In classical physics terms, state A causes state B; that is, the time sequence is that B follows A. This approach assumes a certainty. If A has occurred, then B must follow. In quantum physics, this certainty does not exist; instead, the procedure becomes a question: if you assume state A, is B the subsequent state? At this juncture, Wolf emphasizes that the event is not part of our physical reality as yet. It is an idea waiting for realization. Perhaps Seth would say that we are in the stage of mental activity preceding the creation of the event.

We must now present a short overview of the methods used by the quantum physicist in approaching this problem: will a measuring instrument register state B if the physicist can experimentally prepare state A? First, the complex wave function for the situation is determined from the Schrödinger equation. Then the energy attribute for state B is measured by using an energy operator (mathematically) to decompose the complex wave function into a series of complex wave functions where each element of the series represents a possibility that the energy will turn out to be a given value. This, of course, includes the possibility of attaining energy state B. The probability for an energy state B showing up on the physicist's instrument is computed by multiplying the complex function for B by its complex conjugate.[29] This results in a real number for the probability.

The decomposition of the complex wave function for state A, then, results in all the possible outcomes for this experiment, plus the probability for a given outcome to occur. This condition might be related to the ego having the choice of a myriad events to bring into reality.

Wolf's ideas can be summarized in the following way. Since state A represents a superposition of all possible subsequent events, how is wave B selected from all those choices? It is assumed that the wave function for A flows forward in time and is a complex function. The *mind* then projects a second wave function back in time, which is the complex conjugate of the wave function representing state B. This is analogous to the release of EE units from a gestalt of consciousness. The complex conjugate wave function of B interferes with the function of state A and brings B into reality. It is at this point that the event enters the three-dimensional world by registering on the measuring equipment. Finally, Wolf inserts the mind as the creator of the advanced wave (second wave going backward in time). In a metaphorical sense this is similar to the projection of EE units from a consciousness, all taking place outside our three-dimensional world.

One concept should be emphasized: the EE units are natural emanations from *all* consciousness. Since elementary particles are nonsentient forms of consciousness, they also emit EE units. The conscious aspect of matter is *not* in the three-dimensional world. Therefore, the consciousness portion of an electron can emit EE units and, with the aid of the vast energy pool, can actually create the structure of an elementary particle (electron). The consciousness involved is not necessarily *sentient*. This intriguing notion of a pervasive consciousness is discussed more fully in Chapter 6.

ELECTROMAGNETIC RADIATION AS
AN ASPECT OF EE UNITS

Seth's concept of EE units can be further explained by considering its relation to electromagnetic radiation. Electromagnetic energy is the energy propagated by a vibrating electric and magnetic field created by a charge. Since the field is vibrating, it has a frequency (cycles per second). In fact, there exists a whole spectrum of electromagnetic fields, from gamma rays to radio waves, each differing in frequency, but all traveling at the speed of light. What we call visible light is only a small portion of the spectrum of radiation, though the entire range can be thought of as a spectrum of light. There are no clear boundaries between the various radiations. Einstein's special theory of relativity states that mass is a form of energy, that conversely energy has mass, and that the two are related by the speed of light squared. All radiation frequencies travel at the speed of light, a limiting velocity in our three-dimensional world. In regard to this, Seth says that there are many gradations of matter, or form, that we do not perceive, and that "many of these particles making up such constructions do move faster than the speed of *your* light."

> Your light, again, represents only a portion of an even larger spectrum than that of which you know; and when your scientists study its properties, they can only investigate light as it intrudes into the three-dimensional system. The same of course applies to a study of the structure of matter or form.
>
> There are indeed universes composed of such faster-than-light particles. Some of these in your terms share the same space as your own universe. You simply would not perceive such particles as mass. When these particles are slowed down sufficiently, you do experience them as matter.[30]

This suggests that our electromagnetic radiation spectrum is part of a still larger spectrum. Because our velocities are limited to the speed of light, we are aware of only a small portion of the whole.

Seth's EE units, then, are in effect light units that intrude into our electromagnetic spectrum by slowing their velocity to that of our light and become trapped at that velocity (see Chapter 5 for more detail).

PROBABLE REALITIES:
A QUANTUM PHYSICS CONCEPT

Seth's idea of probable systems expands and enriches his concept of reality. Probable systems may not appear so strange in the light of Hugh Everett's many-universe theory. In attempting to explain the measurement problem in physics, Everett assumed that the universe splits into all its probabilities when a measurement is made. These universes exist side by side and simultaneously, though not in contact. In a similar effort to solve the measurement problem, Eugene Wigner introduced consciousness as the agent that makes the choice between probabilities. In his view, the non-chosen are never actualized. Hiley assigns the task of choice to the quantum potential because he does not want to introduce human consciousness at the level of physics (although it is obviously introduced at a deeper level). With all this as a backdrop, Seth's concepts of probability seem less farfetched.

According to Seth, all actions are initially mental actions, and all mental actions are valid. In our daily lives at any given moment, we have many choices of action, trivial or significant. We choose one of the actions and (in Seth's view) actualize it into our reality by the methods previously described. What happens to the actions we do not choose? He says they are as valid as the ones chosen and are not discarded merely because they were not brought into our universe. They are actualized elsewhere, in other probable realities. That means that in other realities, there are other probable selves, other probable earths, all using actions that we, in this reality, have chosen not to use. In short, our slightest thought gives birth to worlds.

In physics, according to the Copenhagen view, an experimental arrangement brings a particular particle from a state of potential into reality. In Seth's view, where thoughts bring matter into existence, each thought produces worlds merely by the process of selection. Not only do we produce worlds with our thoughts, but each of these new worlds is producing worlds. What is the connection between these realities? Here Seth and Everett part company.

Everett feels that we are unable to perceive other universes and that one reality does not affect the other. But in some sense this is not entirely correct, as demonstrated by the two-slit experiment, the quintessential model for displaying the wave and particle nature of electrons and photons. In the traditional two-slit experiment, electrons are shot at a screen with two slits in it, and the result is recorded on a photographic plate that the electrons strike on the other side of the screen. We have no way of knowing which slit each electron passed through. However, an interference pattern is recorded as if an electron passed through both slits simultaneously. Even when the intensity of the source is reduced to the point that only one electron at a time enters the apparatus, after enough electrons have entered, an interference pattern still appears on a screen. That is, each electron strikes the screen as a single particle, yet the pattern depends on all the paths the electron might have taken. The electron could go through either slit, but as this experiment demonstrates, the path *not* taken affects the result. (Hiley feels that the information from the other probability is effectively randomized or destroyed as far as our universe is concerned.)

Seth disagrees with Everett's conclusion. He suggests that probabilities are an ever-present portion of our invisible psychological environment and that there are profound interconnections binding the probabilities to each other. This connection takes place beneath normal consciousness through a telepathic framework. The inner ego determines which information comes into our conscious minds. Although we do receive information from other probabilities, we choose on an unconscious level which probability to activate in our reality. We have freedom of choice — as do the other probable selves.

Seth says that in dreams and in other situations where the ego is calmed, considerable communication takes place between various probable selves. Because of this cosmic interconnection, we can avail ourselves to some extent of the abilities and knowledge possessed by other probable portions of our personality. In short, we are serially and three-dimensionally oriented, and that prevents us from perceiving our existence in a sea of probable events. As Seth puts it,

> It is impossible to separate one physical event from the probable events, for these are all dimensions of one action.[31]

and,

> All of the probable events of your life exist at once . . . . Since your activities physically must be fitted into a space-time framework, only a minimum of those probable events will physically occur.[32]

In connection with Seth's assertion regarding the inability to separate the probable events from the "real" event, recall Richard Feynman's concept of the sum-over-histories. He stated that when an electron travels from point A to B, it goes through *all* possible paths. The "actual" path is the result of summing all the contributions from all paths. In this way, all probable paths are aspects of the actual path taken. Perhaps Seth is saying something similar.

Seth's statements are also comparable to the views of mathematician Ron Atkin. In his book *Multidimensional Man*, Atkin says:

> Perhaps we can surmise that *the description of a set of events by way of assigning them probabilities is a characteristic indication that the "seeing-agent"* (or 'process of observation') *is traffic of too low a dimension on the action event(s)* — otherwise it would provide us with that single certain event of which these "probable" events are only (geometrical) faces.[33]

He goes on to use the analogy of throwing a die. By observing a single face, the one facing upward, we limit the dimensionality of the event. Through mutual agreement of the parties involved, the dimensions (all faces) are restricted to one. As a result, we must resort to a description of probability or chance. Applying this concept to contemporary physics, Atkin asks, "Have the physicists drifted into methods of observations (or events) which constitute inadequate traffic on a structure which they therefore cannot hope to observe?"[34]

The concept of an almost infinite number of probable selves in an almost infinite number of universes can make us feel rather insignificant. Seth observes that, on the contrary, even though our greater identity is in touch with and aware of all our probable selves, our uniqueness in our universe is extremely important. We seek out and view experience with our own special affinities, and the other probable selves do not deny the validity of our experience or individuality. We are secure because we are choosing from unpredictable fields of actuality those that suit our own particular nature. The ego, or sense of selfhood, jumps in leapfrog fashion over events it does not wish to actualize, leaving other portions of ourselves to choose the events we have *not* chosen.

PULSATIONS, OR ENFOLDING AND UNFOLDING

Another interesting idea of Seth's is that atoms and molecules have a pulsating nature. This observation is not too far out of line from what is understood in physics today. We know from the theory of special relativity that the mass of a particle is actually energy. Planck worked out the relationship between energy and frequency, stating that energy implies frequency, and frequency implies wave motion. Louis de Broglie, using this line of reasoning, clarified the wave nature of matter and postulated its pulsating feature, which was proved in the laboratory by C. J. Davidson and L. H. Germer.

Nick Herbert describes this pulsating nature in the microscopic structure of a coffee cup:

> It is an assembly of *events* rather than of *things*. These events (called *quanta*) last for only an instant, then fade away. Imagine a trillion trillion fireflies flashing in the space of your coffee cup. The cup is a never-still scintillating network of quantum events . . . it is full of *dots*, and the dots are constantly changing. The old-fashioned notion of the cup as made up of atoms is just one frozen frame of the microscopic light show.[35]

In a similar vein, Bohm says:

> The implicate order can be thought of as a ground beyond time, a totality out of which each moment is projected into the explicate order. For every moment that is projected out into the explicate there would be another movement in which that moment would be injected or "introjected" back into the implicate order.[36]

If matter is made up of oscillating fields, what vibrates? Bohm says, space itself.

Seth says that physical matter pulsates. All particles are a series of pulsations at a rate too fast for our biological system to track. We perceive matter as permanent only because our perceptive mechanisms are not equipped to allow us to see otherwise. Particles move in and out with fluctuations that are highly predictable in pattern and rhythm. The integrity of the solid object is maintained because of the consistency of the fluctuations. The particle is a projection into three-dimensional space; the pattern itself exists outside space-time. As the pulsations phase in and out, the particle is actually in our system of reality only at certain points. That is, the particle projects in and out of three-dimensional space at such a rapid rate that we are not aware of the gaps when the particle is not present, so its existence appears continuous to us. During the gaps, the particle exists in another system of reality. In fact, according to Seth, at many points of the cycle, the particle exists in many different universes.

In a similar manner, our psyche or outer ego is being freshly created "at every point" of our existence. Seth says:

> The physically oriented consciousness, responding to one phase of the atom's activity, comes alive and awake to its particular existence, but in between are other fluctuations in which consciousness is focused upon entirely different systems of reality . . . .[37]

Our consciousness pulsates in the same way as matter. Although our outer ego is aware of time only in the three-dimensional universe, the inner ego remembers the periods when we are out of our physical reality. As Wilber says about the moment-to-moment involution of the spectrum of consciousness he calls "microgeny," in each moment we each pass through the entire sequence of levels, from ultimate to causal to subtle to mental to gross, remembering that experience only to the extent that we have evolved.

Since particles of matter are pulsating in and out of the three-dimensional universe, we see them only during a portion of their vibrating cycle. Therefore, we see only those characteristics of the particle that exist in our reality. The total reality is not available to us. Matter, in this view, is not continuous and does not age. Matter is only cohesive enough to give the appearance of permanence. Seth says:

> Matter is continually created, but no particular object is in itself continuous. There is not, for example, one physical object that deteriorates with age. There are instead continuous creations of psychic energy into a physical pattern that appears to hold a more or less rigid appearance.
>
> No particular object "exists long enough" as an indivisible, rigid, or identical thing to change with age. The energy behind it weakens. The physical pattern therefore blurs. After a certain point each

re-creation becomes less perfect from your standpoint. After many such re-creations that have been unperceived by you, then you notice a difference and assume that a change . . . has occurred. The actual material that seems to make up the object has completely disappeared many times, and the pattern has been completely filled again with new matter . . . .[38]

Note the similarity to Nick Herbert's description of the coffee cup.

Seth states that it is through the creation of matter that consciousness can be effective in our three-dimensional reality. The particle is created by CUs approaching our world and then expressing themselves in material form. But matter itself is a pattern of pulsations of psychic energy, creating the illusion of solid objects in three-dimensional space. The physical universe is "being created at all of its points at each moment."

THE ORIGIN OF THE UNIVERSE
AS CONTINUOUS CREATION

Seth has very definite views on the beginning of our universe. These may seem less bizarre if we understand John Wheeler's concept of the participatory universe.

As interpreted by the Copenhagen view, quantum theory tells us that no quantum event is "real" until it is recorded. That is, it exists in *potentia* (using Heisenberg's term) until an observer comes along and brings it into reality. Using this concept, Wheeler proposed a thought experiment that creates an additional paradox.

In a revision of the two-slit experiment (see p. 134) Wheeler introduced a device that combines the slits with a lens that focuses the electrons. A second lens replaces the photographic plate so that the electrons then diverge. If we shoot one electron at a time, the path of the electron is unambiguous, and we know which slit the electron went through. Also, there is no interference pattern since we do not have a superposition of states.

Now, suppose that there is placed in front of the second lens a photographic plate constructed like a venetian blind. It acts as a plate when the slats are closed and appears to be absent when the slats are open. This experimental setup, then, can produce an interference pattern when the blind is closed or unambiguous paths for the electrons when the blind is open. A timing device can open and close the blind at arbitrarily small time intervals. If the electron has already passed through the first lens before we open or close the blind, we can, in a sense, go back and rewrite history. That is, after the electron has gone through the first lens, we can determine whether it is to go through one slit or both. If we assume that the macro world is made up of a large number of micro events determined by the observer, the "real" world is constructed of observer-created reality from an "unreal" micro world. In Wheeler's words:

> Of all the features of the "act of creation" that is the elementary quantum phenomenon, the most startling is that seen in the delayed-choice experiment. It reaches back into the past in apparent opposition to the normal order of time. The distance of travel in a laboratory split-beam experiment might be thirty meters and the time a tenth of a microsecond; but the distance could as well have been billions of light years and the time billions of years. Thus the observing device in the here and now, according to its last minute setting one way or the other, has an irretrievable consequence for what one has the right to say about a photon that was given out long before there was any life in the universe.[39]

Wheeler's delayed-choice thought experiment was conducted in the laboratory and his conclusions confirmed when physicists (Alley, Jakubowicz, and Wickes) at the University of Maryland manipulated a single photon to go through two well-separated paths. By electronically inserting mirrors, they made the photon interfere with itself. Furthermore, the insertion could take place *after* the photon entered the setup. Thus the experimenters could "choose" the photon's past: it could go through one path or both, after the fact.

Wheeler then applied this thought experiment to the big bang version of the universe. The big bang theory says in essence that all matter and radiation were created at some finite time in the past. The universe started out from some initial state with tremendous density and temperature and then expanded and cooled. After some period of time, human beings appeared who could observe the universe. These beings, through the delayed-choice approach, made reality tangible by literally creating the universe, going all the way back to the initial state. Therefore, the universe is not "real" in our terms until we arrive on the scene. There is no big bang — or any other event — until we come along to bring it into reality.

Astronomer John Barrow sums up this point of view in his book *The World Within the World*:

> A most amazing facet of observer-created reality was suggested by John Wheeler. He points out that all the astronomical observations that we make involve the present-day reception of light rays which emanated from distant stars and galaxies billions of years ago. The sea of low-temperature microwaves we observe all over the sky are the dying embers of the Big Bang from which our expanding Universe sprang. At the most fundamental level, all the photons we detect from distant stars are quantum waves whose wave functions are collapsed to classical certainty by the detectors and the astronomers who observe them. Does this mean that in some sense *we* bring these astronomical objects, and even the whole Universe, into being when we observe them today?[40]

Robert Jahn, who conducts research in the fields of both engineering and parapsychology, poses the same question:

> May we not now reasonably ponder that if this powerful space- and time-bending property we call "mass" ultimately traces down to "particles" that we can only experience rather than observe, that we must treat as events rather than as substance, and that we can describe only in quasi-poetic, anthropomorphic terms, then it may indeed be consciousness that establishes physical reality, even on the cosmological scale? Are our minds being stretched, or are our minds doing the stretching? Perhaps another step needs to be taken along the road of scientific conceptualization that has brought us from causal particulate mechanics, to intangible wave/particle dualism, to probabilistic determinism, to observational realism. It is the step that explicitly acknowledges consciousness as constructor of the same reality it perceives, ponders, and postulates.[41]

To understand Seth's views on the beginning of the universe, we must make two assumptions. One, there is an invisible universe out of which the visible universe springs. Two, time is simultaneous; that is, it is from the present that we form our past and future, both collectively and individually. We are not at the mercy of a past we do not control. If this is difficult to accept, give some thought to Wheeler's participatory universe. Perhaps, as we know from quantum theory, each action in the future can be represented by a series of probabilities. One probability is selected, and that becomes our future. Seth says that we can look at the past in a similar manner. There are probable pasts just as there are probable futures. We select only one version of events as our past and ignore the others. Wheeler comments:

> It is wrong to think of that past as "already existing" in all detail. The "past" is theory. The past has no existence except as it is recorded in the present. By deciding what questions our quantum registering equipment shall put in the present we have an undeniable choice in what we have the right to say about the past.[42]

William Irwin Thompson, author of numerous books dealing with the shift to a new paradigm, comes to a similar conclusion:

> All descriptions *of* the past are *in* the present; therefore, history tells our descendants more about us than it does about the imaginary creatures we like to call our ancestors. Like an image before us in the rear-view mirror of a car, the picture of where we have been keeps changing as we move forward in space and time. The narratives of the past from even so short a time ago as the beginning of our own twentieth century now no longer describe us, and so each generation must reinvent the past to make it correspond to its sense of the present.[43]

In the beginning, Seth-wise, there was no God the Father, Allah, Brahma, or even a singularity. In fact, there is no beginning. Instead, we have quite literally the implicate order, a psychological gestalt that exists in what Seth calls the moment point (present moment). The universe is always coming into existence at each moment, and with it comes its own built-in past. (Remember the pulsation concept.) We, in turn, accept only a small portion of the available data that composes each moment. This acceptable data usually corresponds to our ideas of science, history, religious and philosophical beliefs, and other basic knowledge. In this view, our world did not form from mindless atoms coalescing out of brainless gases. Furthermore, it was not created by some distant God on a cosmic assembly line. Rather, it was created from inside out. There was consciousness *before* there was matter. The *potentia* (á la Heisenberg) was there first; the participatory universe (á la Wheeler) brought it into existence. Seth feels that we err in separating matter from consciousness. In his words, "consciousness materializes as matter in physical life." Our own consciousness and our collective consciousness are both part of the divine gestalt.

To clarify this view, Seth asks us to visualize the origin of the universe as similar to the way an image might be created in a painting. Rather than starting with a blank canvas and depicting an explosion, then painting a cell, then a number of cells combining together to form a simple organism, the artist might start with an underpainting, which we may think of as a panel of light. All the world's features and organisms would be sketched in initially, not painted in detail; they would exist in some potential form. Then the panel of light through its own creativity would enrich the colors and delineate each species, plus start the winds blowing and the seas moving. All this activity and energy comes from the underpainting. Seth says:

> I certainly realize that this is hardly a scientific statement — yet the moment that All That Is conceived of a physical universe it was invisibly created, endowed with creativity, and bound to emerge.[44]

Seth emphasizes the fact that *all* portions of the universe are conscious, with the planner within the plan.

> There is of course no "outside" into which the invisible universe materialized, since all does indeed exist in a mental, psychic, or spiritual realm quite impossible to describe. To you your universe seems, now, objective and real, and it seems to you that at one time at least this was not the case, so you ask about its creation and the evolution of the species. My answer has been couched in the terms in which the question is generally asked.

> While you believe in and experience the passage of time, then such questions will naturally occur to you, and in that fashion. Within that framework they make sense. When you begin to question the nature of time itself, then the "when" of the universe is beside the point.[45]

The facts we accumulate in science (and elsewhere) can be viewed as water poured into a container, and the container represents our concept of time. Our facts take on the form of the container, and all of our theories are couched in terms of the container. If we were to re-place the container with one in the form of no-time, such concepts as the big bang and evolution would be viewed differently. Alex Comfort expresses it this way:

> When we extrapolate backwards, whether to the cosmological Big Bang or to a landscape inhabited by saber-toothed tigers, our assumption is that things (these things) were there before we were. If we express it in space-time, reality "was there before" we were, and fossil and extinct organisms may have been observed as such by other animals having comparable ways of abstracting reality to ours, but the human type of reality is contingent on the presence of humans to observe it. It is arguable that in the absence of human brains, these things "were not there" though the materials for their construction were inferentially there if such brains had been represented at that point in space-time. This bit of hair-splitting is important, because it bears heavily on some of our formalisms for evolution, organic or cosmological, in which linear historicity is tacitly assumed.[46]

Bohm's view of the beginning is similar to Seth's except that it is a timed order. Bohm sees the universe as a little ripple arising from the vast ocean of energy (or plenum). The ocean of energy is the implicate order. In a similar manner, Seth sees the universe as springing from another, deeper level he calls Framework 2 (discussed in the following section). He states that our visible universe results from EE units from the invisible universe becoming manifest. Here is Bohm's description of the beginning:

> This plenum is . . . no longer to be conceived through the idea of a simple material medium, such as an ether, which would be regarded as existing and moving only in a three-dimensional space. Rather, one is to begin with the holomovement, in which there is the immense "sea" of energy described earlier. This sea is to be understood in terms of a multidimensional implicate order . . . while the entire universe of matter as we generally observe it is to be treated as a comparatively small pattern of excitation. This excitation pattern is relatively autonomous and gives rise to approximately recurrent, stable and separable projections into a three-dimensional explicate order of manifestation, which is more or less equivalent to that of space as we commonly experience it.[47]

In Bohm's view, the big bang is expressed as a ripple on the vast holomovement. He uses an imaginative analogy to explain it:

> An interesting image is obtained by considering that in the middle of the actual ocean (i.e., on the surface of the Earth) myriads of small waves occasionally come together fortuitously with such phase relationships that they end up in a certain small region of space, suddenly to produce a very high wave which just appears as if from nowhere and out of nothing. Perhaps something like this could happen in the immense ocean of cosmic energy, creating a sudden wave pulse, from which our "universe" would be born. This pulse would explode outward and break up into smaller ripples that spread yet further outward to constitute our "expanding universe". The latter would have its "space" enfolded within it as a special distinguished explicate and manifest order.[48]

So, Wheeler posits a participatory universe that comes into existence through the presence of an observer; the universe becomes "real" only after we arrive on the scene. Seth says our universe springs from a "divine psychological gestalt" which exists in the moment point. Bohm says the universe is a ripple on a "sea of cosmic energy," the holomovement. All require that a choice be made. All have a director of the action who makes the choice — whether it is Wheeler's observer, Bohm's superimplicate order, or Seth's inner ego, which, like Wilber's self, manipulates the three-dimensional world.

FRAMEWORKS OF EXISTENCE

Seth, like Wilber and Bohm, talks about a spectrum of consciousness, but in terms of frameworks rather than orders or levels. He gives detailed explanations for Frameworks 1 and 2, mentions Frameworks 3 and 4, and implies an unending order of additional frameworks, "composed of ever-spiralling states of existence . . . ."[49] All these frameworks exist one within the other, and each impinges upon the other. The fact that Seth refers to them separately does not imply a spatial separation; to some extent, all realities are immersed in all others. Even the highest frameworks are not inaccessible; the more our thoughts aspire toward them, the more likely we are to reach them.

Our discussion will be confined largely to Frameworks 1 and 2, for which we have the most information. Seth explains these frame-works by asking us to imagine that we are seated in front of a televi-sion set with a remote control that has access to many channels. The physical events of our lives are analogous to the events flashing on the screen as we switch from one channel to the other. Our three-dimensional world in which these events take place is Framework 1. Watching a program, we are not aware of the activities in the televi-sion studio that make the production possible. The realm of the stu-dio, which is the creative background for the program we watch, is like Framework 2.

The first point made with this analogy is that just as we choose which channel we watch and thus bring it into our world as a physi-cal event, so do we choose the events of our lives. We are not com-pelled to choose any particular channel. We can ignore those chan-nels that offend or do not interest us and watch only those programs that appeal to us or serve our purposes. The television channels sim-ply offer us those events that are possible, and we make the choice.

Second, our everyday experience (the program we are viewing) represents a joint effort between the viewer of the programs and the persons responsible for its production, who are in Framework 2 (the television studio). If we assume that the television programming re-flects the "overall, generalized emotional and psychic patterns of the age," as Seth says, then it does not *cause* the events of our lives; rather it casts back to us events of our own (unconscious) design. When the program displays violence and greed, the writers and producers are working with material from the collective unconscious, and the val-ues of our culture are reflected back to us.

Each actor in the television dramas, and for that matter each ob-server, chooses his/her role. Just as newspapers and magazines in-form us which programs are to be shown, so do our inner selves have information about issues raised throughout the world. We decide which channel to watch, or which life experiences we wish to participate in. As an example, supposing someone feels a need to spend the remain-der of his life behind prison bars for murder, someone else, for their own purposes may volunteer to be the victim.

Framework 2 is the creative source of our three-dimensional world, and we rely on it to help us through our daily routine. Seth says:

There is a program director, who must take care of the entire programming. Shows must be done on time, actors assigned their roles. Our hypothetical director will know which actors are free, which actors prefer character roles, which ones are heroes or heroines, and which smiling Don Juan always gets the girl — and in general who plays the good guys and the bad guys.[50]

The program director is our inner ego. The only job of the outer ego is to choose the programs. Of course, other outer egos are choosing and viewing the same programs, but each reacts as an individual. The background activity of the inner ego is largely unknown to our outer ego, but each event in our everyday reality is created by and dependent upon these inner mechanisms. "The vast mental studio of Framework 2" is the source of all our experiences of everyday life.

It is important to note that each of us is sitting in front of his/her own television screen. Furthermore, the screen for each individual is correlated and coordinated with all others so that we can intelligently engage in and converse about the same programs. In essence, we each have our own three-dimensional environment. This concept will be discussed in more detail in Chapter 7.

Through our conscious level alone, we could not keep our bodies alive, yet life seems to flow through us automatically and spontaneously. Our breathing, circulation, digestion, elimination, and all other physiological processes, along with maintenance of our psychological continuity, are controlled in Framework 2. Seth sums up his description, still using television as an analogy, as follows:

Actors visit casting agencies so that they know what plays need their services. In your dreams you visit "casting agencies." You are aware of the various plays being considered for physical production. In the dream state, then, often you familiarize yourself with dramas that are of a probable nature. If enough interest is shown, if enough actors apply, if enough resources are accumulated, the play will go on. When you are in other than your normally conscious state, you visit that creative inner agency in which all physical productions must have their beginning. You meet with others, who for their own reasons are interested in the same kind of drama. Following our analogy, the technicians, the actors, the writers all assemble — only in this case the result will be a *live event* rather than a televised one. There are disaster films being planned, educational programs, religious dramas. All of these will be encountered in full-blown physical reality.[51]

Since all these real-life events are the result of individual choice, nothing occurs by chance, including birth and death. Framework 2 is a huge creative atmosphere where all events are known everywhere. The communication network is instant, and each mental act makes its imprint on the nearly infinite multidimensional screen of Framework 2. We tap into this vast storehouse in many ways, particularly through dreaming. In the dream state, all probable events are available, just as the physical events of our daily lives are available to us during waking states.

Framework 2 can be seen as an infinite information service. It provides us instantly with any knowledge we deem necessary. It computes probabilities in a flash. But it is *not* a cold, mechanical supercomputer; it has loving intent; it has everyone's best interest in mind. That means that Framework 2 cannot be used to force a particular event on someone else. Your desires cannot conflict with those of another being and be physically actualized. Seth says that All That Is — the Absolute, the holomovement — is a constant presence that surrounds and engulfs us. It is a well-intentioned, gentle, powerful, and all-knowing atmosphere, and each individual is a manifestation of this presence.

Obviously, if we are dividing reality into two frameworks, the inner reality can be correlated with Framework 2 (and beyond) and the outer reality with Framework 1. We see our outer reality as concrete, factual, definite; events and facts make up our daily lives. These events and facts rest upon scientific data that, at least in our view, are unassailable. On the other hand, the fantasies and myths of the present and past are often regarded as unreal or imaginary. One of the reasons we do not view myth as a kind of truth is that science (at least Newtonian physics) does not model reality beyond Framework 1.

The discovery of the subatomic world and the subsequent development of the theory of quantum mechanics forced scientists to abandon the classical Newtonian model with its picturable and commonsense metaphors. When physicists were faced with the complex and paradoxical problems of the micro world, they turned to mathematics, and the general population tuned out. Now even mathematical modeling is proving extremely difficult in the realm of the subquantum.

In the past, the task of modeling counterintuitive events belonged to myth, not to mathematics. Myth, however, has been denigrated because it does not have the precision we have come to expect in our highly technological world. Actually, myth is often more truthful than so-called "facts," but the myths are not to be taken literally. Myths are malleable enough to represent the ineffable. They can be stretched and modified to provide useful containers for society's viewpoint.

A myth can achieve the status of truth when it illustrates primordial reality (Framework 2) — the time before time. The myth reflects first principles through an analogy to an event that cannot be imagined in terms of our three-dimensional world. The myth describes reality like an artist unfolding meaning through a work of art, not like an engineer explaining how something works. In a sense, the myth is an attempt to describe the infinite in finite terms.

To Seth, myth is the womb out of which fact must come, the inner reality from which our world springs. Thus, the making of myths is a natural psychic phenomenon. We program our lives on the basis of the myths that are prevalent in our society. What we describe as fact is sifted through this strainer of our societal myths to determine if it is indeed correct. Or we may say that myth is a lens through which all experience is received.

A myth, when accepted by a culture, is considered fact. All societies have myths, but they are not always recognized as such. Seth comments on both religion and science as myths:

> In those terms, then, Christianity and your other religions are myths, rising in response to an inner knowledge that is too vast to be clothed by facts alone. In those terms also, your science is also quite mythical in nature. This may be more difficult for some of you to perceive, since it appears to work so well. Others will be willing enough to see science in its mythical characteristics, but will be most reluctant to see religion as you know it in the same light. To some extent or another, however, all of these ideas program your interpretation of events.[52]

Seth goes on to say that danger signals should go up when a myth represents itself as fact: "Facts are very handy but a weak brew of reality. They immediately consign certain kinds of experiences as real and others as not."

Framework 2, then, is an everchanging interior world housing the "dreams, hopes, aspirations, and fears as well as the emotional reality of all consciousness." It is in a constant state of flux which, when manifested, forms the events of our physical reality. "All consciousness" to Seth is indeed all-inclusive. Not only is he referring to amoebae, elephants, and rose bushes, but also to the earth and the stars and elementary particles. Each organizes and interprets the world according to its own characteristics and, in the case of human beings, according to their mythic outlook. It is therefore possible to fashion worlds from perspectives unlike the one familiar to us. One view is not superior to the other; rather, each culture represents a particular focus shaped by its own mythical structure. Each culture adopts an official psychological outlook; each civilization molds Framework 2 in a particular way.

In addition to forming models and patterns, Framework 2 acts as a source of creativity for our outer reality. Seth says that if an artist operates exclusively from belief in laws of cause and effect, the artistic endeavor will be bound by the memories and experiences of Framework 1. That is, only that which fits the particular myths subscribed to in outer reality will be drawn from the artist's inner reality. Restricting the use of that tremendous creative pool of Framework 2 results in works that fail to touch us deeply. (An example is the social realist art under Stalin's regime.) Such restriction also applies to our everyday activity; we can live according to the barriers of cause and effect and the myths of Framework 1, or we can tap the pool of Framework 2 in which barriers and compartments do not exist. By utilizing the creative pool, we are open to flashes of insight or (following Bohm) to meanings that come from indefinite depths of implication.

Seth points out that our present view of reality is based on the assumption that we have a finite amount of usable energy available in our universe (second law of thermodynamics), and that a given amount of energy will produce a corresponding amount of work. In other words, we assume that Framework 1 is all there is, that Framework 2 does not exist, and that as the original source of the big bang slowly wears itself down, no new energy is inserted into our universe. By postulating Framework 2 (and frameworks indefinitely beyond), however, Seth sees energy as being inserted continuously into Framework 1. Through the pulsations described by Bohm as enfolding and unfolding and by Wilber as microgeny, our psyches and matter are constantly rejuvenated from other frameworks. In a sense, we are perpetually being reborn, as is every particle in the universe. Again, as with Bohm's spectrum of orders, there is no closed system.

Seth touches on an additional aspect of Framework 2 that has relevance to our discussion. Framework 2 is the medium in which our universe exists both physically and psychically. Throughout history individuals have had glimpses of this inner reality through dreams, meditation, and altered states of consciousness. Seth says these glimpses are actually brief visits to Framework 2. The individuals involved think of their glimpses as representative samples of the whole. Such assumptions can be erroneous, resulting in distortions that are often incorporated into the myths accepted by a particular society.

An example Seth gives of such distortion is in the writings of Plato. Seth says that Plato saw Framework 2 as a realm that is eternal and unchanging, a perfect but frozen composite, which is represented by Plato's theory of forms. On the one hand, this vision inspired people toward perfection, but it also reproached them for falling short of absolute faultlessness. The problem was that in this view human beings could not change these ideals, but could only aspire to them.

Descriptions of inner reality require symbols to translate ineffable experience into outer reality. The symbols often become hardened into fact; the map is mistaken for the territory. Even the Eastern religions, in Seth's view, have some distortions regarding Framework 2. To quote:

> You are not fated to dissolve into All That Is. The aspects of your personality as you presently understand them will be retained. All That Is is the creator of individuality, not the means of its destruction.
>
> My own "previous" personalities are not dissolved into me any more than your "past" personalities. All are living and vital. All go their own way. Your "future" personalities are as real as your past ones.

> After a while, this will no longer concern you. Out of the reincarnational framework, there is no death as you think of it.[53]

This idea is certainly in contrast to the Perennial Philosophy. More discussion on these differences will follow.

Framework 1 knowledge seems to come from two sources. First, we have that body of facts accumulated by humankind during the course of time. This information has been transmitted through verbal means, records, books, and even television. It has been processed through our societal myths and then established as facts. This is "systematized" knowledge; each discipline has organized its information according to its own model. Second, there is the information that comes through our senses. But as Seth repeatedly indicates, this also is filtered through our own beliefs, wishes, and purposes. Both these forms of information constitute what we might call physically available knowledge. But if we assume that Framework 1 is not a closed system, then our physically available facts from Framework 1 are restricted. Framework 2, on the other hand, is an infinite storehouse of knowledge; Framework 2 and the inner ego provide far more than seems physically available. As Seth says:

> I do not want to compare the inner ego with a computer in any way, for a computer is not creative, nor is it alive. You think of course of the life that you know as LIFE, in capitals. It is, however, only the manifestation of what in those terms can only be called the greater life out of which your life springs. This is not to compare the reality that you know in derogative terms to the other-source existence, either, for your own world contains, as each other world does, a uniqueness and an originality that in those terms exist nowhere else — for no world or existence is *like* any other.[54]

What Seth is saying is that in each now-moment (what we call the "present" moment, which implies past and future), we draw from an enormous bank of unpredictable actions in Framework 2 those actions that are significant to us. These are expressed through myth and result in what seems to be predictable behavior.

> Propensity is a selection of significance, an inclination toward the formation of selected experience. This applies on all levels — atomic and psychological — and to biological stimulus and mental intent.[55]

In conclusion, Framework 2 is available to all of us. In fact, we are manifestations of it. Framework 2 is the atmosphere within which we live, a living and loving condition that surrounds us, interpenetrates us, and actually is us. It leans in our direction, but not at the expense of anyone else. We are all interconnected manifestations of the whole, but at the same time each of us is a unique materialization like no other.

DREAMS: AN EDUCATIONAL PROCESS

Another area in which Seth's ideas correspond with those of major thinkers is the significance of dreams. Dreams are sources of divination, healing, reflections on the waking state, and experience of the supernatural. In recent times, dreams have been recognized as the source for creative ideas; artists, writers, and scientists relate that dreams have provided inspiration and even solutions for particular aspects of their work.

In the rationalist societies of the West during the nineteenth century, doctors generally believed that dreams were hallucinations and essentially meaningless. The modern approach to dreams originated with Sigmund Freud, who placed dream study on a scientific footing with the publication in 1900 of his comprehensive analysis, *The Interpretation of Dreams*. A product of his age, Freud was a mechanist. He saw the human brain as analogous to a steam engine and described it in thermodynamic terms. In Freud's view, the brain is a complex network of excitory neurons transmitting impluses to other neurons and producing a build-up that required discharge. The dream was considered a vehicle for this discharge.

Within this concept, Freud gave the dream two specific functions. One was to fulfill repressed unconscious wishes, which were largely sexual or aggressive. The other was to allow the ego, the censor of all those repressed wishes, to get a good night's sleep. A waking thought that cannot be expressed because of social convention runs through the brain building up psychic energy. During sleep such repressed thoughts combine and stage a play in symbolic form — the dream — to discharge the excess energy. The symbols disguise the unacceptable thoughts, which would otherwise disturb our sleep.

Whereas Freud saw the unconscious as a repository of unbridled lust, aggression, and wishes originating in our childhood, Jung offered a different view. He was disenchanted with Freud's belief that psychic energy was primarily sexual and thought that Freud's interpretation of dreams and the corresponding symbolism was much too restrictive. Jung felt that dreams are not just repressed images lodged in memory, but that they actually produce new information for our conscious mind. To Jung, Freud's symbols were just signs; true symbols point to a higher level of consciousness. When the dream symbols were primordial images, Jung labeled them archetypes — inborn thought patterns that originate in the pool of the unconscious common to all humankind.

Archetypes are symbolically represented not for purposes of disguise but to transmit to the conscious mind the information in the collective unconscious. The symbolism is a means of expressing that which is unexplainable. Jung felt that a part of our psyche (which he called the Self) integrates our total psychic life through the medium of the dream. Since the information in the dream is not expressed directly but through symbolism, the dream has to be explored as a whole, and from every angle, to determine its meaning.

To Freud, the purpose of dreams is to permit uninterrupted sleep. To Jung, dreams are the means to obtain the unconscious information we need for a complete psychic life. To Freud, sleeping is primary; to Jung, dreaming is primary. In support of Jung's view, Jeremy Campbell has this to say in his book *Grammatical Man*:

> Dreams are not chaotic, nor are they irrelevant. They supply new context and thus enrich knowledge. By doing so they serve an educational function for the dreamer, because they lead him across a dimly lit landscape of riddling images, from information he knows to information he does not know yet in his conscious, waking life.
>
> Dreams use information as a means of control, helping to put the waking life in order, providing stability, balancing the unbalanced.[56]

In 1953, the study of dreams moved from the psychoanalyst's couch to the laboratory. In that year, at the University of Chicago, Nathaniel Kleitman and Eugene Aserinsky, while studying the eye movements of sleepers, noticed bursts of rapid eye movements (REM) occurring about four to six times during the night. They determined that there is a correlation between REM sleep and dreaming. This discovery brought the investigation of dreams into the laboratory where the physiological aspects of REM sleep have been extensively studied. So far, no one has correlated subject matter of dreams and physiological changes. It is known that deprivation of REM sleep causes abnormal behavior, therefore dreaming is quite necessary. But understanding of the REM state is not sufficient to comprehend the meaning of dreams. Finally, the discovery of REM sleep confirmed that human beings (and, for that matter, animals) dream every night.

Seth agrees with the necessity of dreams and the adverse effects observed when dreaming is curtailed. He comments:

> As is known, anyone deprived of sufficient dreaming will most likely begin to hallucinate while in the waking state, for too much experience has built up that needs processing. There are many secondary hormonal activities that take place in the dream state and at no other time. Even cellular growth and revitalization are accelerated while the body sleeps.[57]

A particularly interesting area of research is the study of lucid dreaming, in which the dreamer is aware of being in a dream state and often is able to consciously control the outcome of dreams. This ability can be taught, and some people have it naturally. The most extensive research on this topic is being done by Stephen LaBerge at the Stanford Medical Center. LaBerge says that when we are fully lucid during dream states, we are in complete control and have full responsibility for the dream's content. Furthermore, LaBerge's subjects are capable of discovering answers to problems from their waking consciousness by reenacting the problem in their dreams and trying out various approaches.

There is a close relationship between the out-of-body experience and the lucid dream. If, in the lucid dream, the dreamer is aware of his/her sleeping body, it is as if he/she is indeed "out of body." Also, there is some evidence that lucid dreaming is a spontaneous meditative state. In the Buddhist tantric tradition, initiates are often taught to enhance their meditative practice by using the lucid dream as a natural form of meditation. Lucid dreaming may be seen as an "advanced" form of dreaming. It has been reported that shamans develop the technique of lucidity in order to enhance their knowledge and thereby increase their healing abilities.

Seth has a great deal to say about dreams.

> When you look into a mirror you see your reflection, but it does not talk back to you. In the dream state you are looking into the mirror of the psyche, so to speak, and seeing the reflections of your own thoughts, fears, and desires.

> Here, however, the "reflections" do indeed speak, and take their own form. In a certain sense they are freewheeling, in that they have their own kind of reality. In the dream state your joys and fears talk back to you, perform, and act out the role in which you have cast them.[58]

The dream, in a symbolic way, plays out our conflicts, reflects our deep feelings, and allows our desires to be symbolically expressed. The entire production involves the evolution of an emotion or belief. In the dream state, we set a thought free and see what happens to it, how it develops, where it goes. Since the dream is, in a sense, a theatrical performance created by us, we can be an observer in the audience, or one of the actors, or we can move in and out of roles at our discretion. All the symbols used are our own private ones; they are our psychic shorthand. These symbols can change. If we could spread out our dreams serially from birth, we would find that we change the meaning of our symbols as we grow, or the content of the dream itself may change the symbols.

In any event, in the dream state we work out problems and challenges. Thus, dreaming is a practical necessity. Seth says that if we really understood that, our dreams would be even more beneficial to us: "Whether or not you remember your dreams, you are *educating* yourself as they happen."[59]

Seth contrasts the waking and dreaming environments by describing the dream as analogous to a stage set. In our waking state, if we wish to detach from a certain event or place, we simply move away from it in our familiar three-dimensional space by walking or driving or flying on an airplane or whatever means of locomotion are available to us. This action takes place only because of our conscious intention. Seth says that the same is true in our dream reality. We do not go anywhere in a spatial sense, but through our unconscious desires and wishes, we can easily change the scenery merely by clearing the old set and installing a new one. The contrast is clear: in three-dimensional space, we seem to move about within an existing environment, whereas in the dream state we create new environments simply by intent. In both cases, Seth says, it is our wish that produces the change.

Comparing an object in our physical world with a thought or emotion in our dream world, Seth asks us to imagine an oak tree in a meadow on a bright, sunny day. It casts a shadow on the surrounding turf. A photographer recording the scene might include the shadow as well as the tree to enhance the quality of the photograph. No one would confuse the shadow with the tree, and no one would say that the shadow is unreal. If one were to visit the meadow on a windy day, the shadow would seem to move in direct relationship to the movements of the tree. Each shadow of each leaf would follow with faithful obedience the motion of the leaf that gave it birth. In the dream world, Seth says, our ideas, thoughts and emotions are the "objects" of our concern. They serve the same purpose as solid objects do in our physical world and are as real in our dreams as oak trees are in our waking life. Thoughts and feelings encountered in dreams also cast shadows of a kind, which Seth calls hallucinations. These too are as real in the dream as the shadow is in the physical world. These hallucinations have a strong role to play in dreams and add immeasurably to a dream's effectiveness, providing texture and quality in a fashion similar to the shadows in a photograph. But unlike the shadow, the hallucination is not passive, not completely subordinate to the originating thought or feeling. Hallucinations are, in a sense, conscious, and display abilities of their own, as if the shadow of the tree, once cast, was free to pursue its own imperatives. A creative exchange exists between the thought and its hallucination.

If we were as conversant with our dreams as we are with our physical world, we would have no difficulty distinguishing between a hallucination and its source. But we are not, and it takes some experience to learn this. Until we learn it, the hallucinations may distort our dreams and make them seem confusing, chaotic, meaningless, or even frightening.

Hallucinations and shadows differ in another important respect. If we were to tell the shadow of an oak tree to go away, it would remain unaffected by our request. But in the dream environment, since the hallucination was created by our own mental activity, we can withdraw the source (the thought or feeling) and the hallucination will disappear. That is, we can literally tell the hallucination to leave, and it will do so.[60]

In the dream state, emotional mood plays a much more obvious role than in the waking state. While the three-dimensional world is created in cooperation with other individuals, the world of dreams is not constrained in this way: each of us is the sole contributor to our dream environments. Thus dark moods turn into dark landscapes; desolate surroundings in our dreams are the shadows cast by our own feelings of desolation. Seth points out, however, that awareness of shadows cast by the mind can have a constructive effect, such as helping us to face our fears more directly. Apparently, the more honest our thoughts and feelings while awake, the less disturbing will be our shadows while asleep.

Dreams can involve not only information processing but also information gathering. Seth tells us that long ago humankind used dreaming as a "science." Dreams were used as a source of visions to locate abundant food supplies, and group dreaming was employed by the tribe in a similar way, for instance, in the midst of a drought to find water. Seth describes this:

> The various tribal members would have dreams in which the problem was considered, each dreamer tackling whatever aspect of the problem that best suited his or her abilities and personal intents. The dreamers would travel out-of-body in various directions to see the extent of drought conditions, and to ascertain the best direction for the tribe to take in any needed migration.[61]

The dreams would be discussed the next morning. In addition, the group might check with a neighboring encampment to discover the nature of *their* dreams. These dreams might be quite direct, or they might be conveyed symbolically. In either case, all members understood that the dream had significance for the group as well as for the individual.

In his book *Synchronicity*, David Peat describes dreaming among the present-day Naskapi Indians.[62] This tribe's use of dreams is strikingly similar to Seth's description of dreaming among early human beings. As our species identified more and more with our exterior environment, this ability to use dreams diminished. Once we lost direct contact with the inner ego, a duality was created: a self who acts in the three-dimensional world and a separate spiritlike soul (inner ego) that operates in a nonmaterial realm.

To summarize Seth's views on dreams:

- The conscious mind is capable of interpreting dream information.
- We are conscious in both our dreams and our waking.
- Future probabilities are worked out through our dreams, for the whole species as well as for the individual.

Regarding the physical basis of dreams, Seth says that dream images are like waking images but are built of particles not visible in the waking state; that is, the EE units are not slowed down enough to be physical matter.

TIME, A FEATURE OF FRAMEWORK 1

The greatest difficulty Seth encounters in describing his version of reality is the need to tailor his explanations to our root assumptions. Root assumptions are our built-in ideas of reality, unquestioned agreements upon which we base our ideas of existence. One of the root assumptions of our three-dimensional universe is the concept of time and space. Seth agrees that this assumption is necessary if we are to experience the life we know in this world. Because we are not yet capable of recognizing mental acts as valid, we must materialize thoughts in order to interpret them. We accept that mind *influences* matter, but we do not yet understand that mind *creates* matter.

Using Seth's terminology, the outer ego understands most events in physical terms, and for a given event creates a past, present, and future. The inner ego exists in the moment point, in which there is no past and future separate from the present. The moment point concept is difficult to visualize, but we can perhaps get a glimmer of understanding through a metaphor. Seth says that all probabilities and all realities flow through the moment point.

For simplicity's sake, let us imagine that we have access to only those probabilities that our level of consciousness can handle. Using Wilber's terminology, we have a basic deep structure open to us, and each moment has access to it. For us the difference between moments is that we select distinct events for each moment and thereby create a time sequence that arises from those distinctions. But suppose we are a consciousness capable of accessing all events at all moments. Then a time sequence is not necessary and all moments can be accessed at one moment. For this consciousness, there is really only one moment, the moment point. Obviously, when a psychological framework is timeless, it is difficult to deal with explanations that require a "beginning." Accounts of the origin of the universe and biological evolution are almost impossible to consider without reference to time.

Seth describes Framework 2 as a large library of knowledge — infinite, in fact. From this library we draw each now-moment that we deem significant to us and our development. Although it is infinite, Framework 2 is not static or complete. Each event we select from a set of probabilities automatically gives rise to new probabilities, thereby continuously increasing the source of knowledge. The selection of a probable action does not lead to a particular inevitable event. Instead, each selection has offshoots in infinite directions, and these in turn have offshoots. The outer ego chooses "from unpredictable fields of actuality those that suit its own particular nature . . . the selfhood [outer ego] jumps in leapfrog fashion over events that it does not want to actualize and does not admit such experience into its selfhood."[63]

In Bohm's description of movement, a wave comes into focus in a small region of space and then disperses. A similar wave comes into focus at a slightly different point in space and then disperses. The focusing and dispersing at different points in space defines a particle's motion. The important point to remember is that the incoming wave brings information from all points in space, while the outgoing wave spreads information to all points in space. A similar analogy can be used to describe Seth's conception of time.

Let us think of Framework 2 as an infinite ocean of all possible moments and events from which the outer ego can actualize individual moments. The moment that is chosen is the now-moment. The conglomeration of moments can be viewed as a superposition of waves representing all events and completely surrounding and available to the now-moment, much as the information from all parts of a holo-gram is available at every point in the hologram. When the outer ego selects an event, filtered through its own needs and affinities, that selection in effect collapses the wave function. New probabilities are then created by the event and flow out into the ocean of Framework 2.

At that moment, how do we view the past and future? Past mo-ments are back in the ocean and no longer have three-dimensional existence. (In physics terminology, we would say that they become part of the superposition of events.) If we need a past at that moment, we construct it by selecting a series of events from Framework 2 that we (individually or as a group) deem probable. In actuality, this past is only one thread of events from the vast fabric of possible pasts. It appears in consecutive order because of our selection. To us, it was the most probable past. But we could all agree on another probable past and just as easily consider that history.

Within a society, different groups can and do choose different pasts. As an example, one group sees Jesus as the son of God who lived 2000 years ago. A second group's history depicts Jesus as a Zealot who stormed the temple in order to expel Roman rule. Another group says that Jesus never existed. Since all probable events are actualized some-where, all three histories can be made acceptable by common consent (though if a past seems highly improbable to the group selecting, con-sensus will not be reached). In this manner, all Everett's parallel uni-verses (see Chapter 1, pp. 33-34) are in existence, and in Seth's view, all are interconnected in Framework 2.

What about the future? This, also, is not determined, just as the past is not determined. As information flows *into* the now-moment, it also flows *out* in terms of probabilities. Seth calls these probabilities "ghost images," and they serve as mental stimuli for exploring the future. But as with the past, any one future is only probable, and the selection process is open at any moment. If these ideas seem counterintuitive and unscientific, note Wolf's comment:

> Mind does not observe the past; it creates it. The past is only as real as we make it. If this is so for the past, what can we say about the present? The present depends on which choices were made (rather, will be made) in the future. I am as I am because I will be as I will

be. I was as I was (in the past) because I am as I am (in the present), not the other way around.[64]

One of the ways to visualize the moment point is through a scheme employed by Ouspensky and described by E. A. Abbott in his book *Flatland*. Abbott devised a fictional universe for an entity called "Square," who exists on a two-dimensional plane of zero thickness. Square does not have thickness and does not understand the concept.

Abbott asks us to assume that Square's third dimension is time instead of space. Imagine a three-dimensional cigar passing through Square's two-dimensional space. The vertical axis represents time; the portion of the cigar below the horizontal line represents the past, the portion above the horizontal line the future, and the horizontal is Square's present. As the cigar moves slowly downward through the time axis, Square notes a change in the circular area defined by the cross section of the cigar. To Square, this Circle appears to pass through a life cycle, beginning as a small ring (the lower end of cigar) and slowly growing to maturity (the middle of cigar). After this point, as more time "passes," Circle becomes smaller and smaller until it disappears or dies.

An observer who can visualize the third dimension of space but not the fourth dimension of time would not see the birth and death of Circle but would view the entire cigar in a timeless present. For this observer, Square's time would be illusory. In physics this is described by the four-dimensional space-time developed by physicist Hermann Minkowski. In his view, the past and future exist together. As Herbert observes, "Minkowskian space-time is a kind of frozen snapshot of eternity."[65]

While the Flatland story gives us some conception of what is meant by a no-time Framework 2, it falls short of Seth's description of probable realities. In Flatland we would have a very deterministic world in which Circle's life cycle would be completely determined from beginning to end. This is certainly not Seth's reality.

Seth uses another image that expands upon the Flatland analogy to bring us closer to his conceptions. In the original story, Circle's past and future are laid out in a determined sequential arrangement. Seth introduces the image of a wagon wheel to suggest that Circle has choice, or input, into the process. The hub of the wagon wheel represents the present moment for Circle. The spokes of the wheel, which are infinite in number, represent probable actions that Circle can "choose" as past and future at any present moment. She may choose a spoke for a future and a spoke for a past. Now, it would seem that by choosing these spokes, Circle is stuck once again with a determined past and future, but such is not the case. Instead of one choice, Circle gets to choose moment by moment. That is, moment by moment as the hub moves along through the present, a whole set of spokes are presented to Circle from which to construct the past and future.

With this revised analogy, we can see that the moment point includes every possible event on every possible spoke for both past and future. All events exist in the moment point, or no-time. To Circle all things are possible, including designing the past at each "now" moment. At each instant, Circle not only chooses moments from the moment point but also creates new ones in the process.

The concept of probable events also can be described by comparing it to that portion of the electromagnetic spectrum we term visible radiation. A prism spreads out the various wavelengths of visible light into a range of hues from red to violet. Since there is not a discrete line between each color, we can postulate that the light actually consists of an infinite number of colors. Similarly, each event has an infinite number of probable events implicit within it. If we choose to live in a certain reality that we call red, we should understand that we are experiencing only one portion of the total: the moment point contains all the realities of the rainbow simultaneously.

If we superimpose this analogy on the Flatland story, the moment point contains all the realities of the rainbow *plus* all the pasts and all the futures with all their probable events. As we proceed up the consciousness scale, or broaden the context by incorporating more dimensions, the event seems to include more of the whole. All the aspects become more unified.

When our analogies are applied to concrete examples, we see that many of our three-dimensional views are somewhat simplistic. One example is the concept of reincarnation. The Eastern approach to reincarnation is that the succession of lives is not random but determined by the law of karma. Karma is an application of causality to moral conduct. In a simplified way, karma can be seen as the collective force developed by a person's past actions. That is, our present character is the result of our past moral conduct and, in turn, determines our future fate. Similarly, in the Judeo-Christian tradition it is said, "Whatsoever you soweth, that shall you also reap." This kind of moral causality implies that every cause has an effect. Such a view, of course, requires sequential time.

How would this be seen through the lens of Seth's moment point? While Seth accepts the Eastern concept of reincarnation, he sees our interpretation of it as distorted because of our need to understand it through a timed order. In the three-dimensional or causal view, if a person lives in five different centuries, it seems to us that the lives in the earlier centuries occur before the later ones and, as a result, the karma built up in the earlier lives affects later lives. In this view, the process is not reversible. It seems ludicrous to think that a twentieth-century event might influence a life led in the fourteenth century, and it seems logical to assume that if the life of a fourteenth-century person is finished, any further development must be accomplished in subsequent times. That is, all of the challenges of the later life are completely dependent on the results of earlier lives. These assumptions, however, have validity only in our timed three-dimensional world. When the process is seen from the moment point, our understanding is quite different.

From the point of view of our multidimensional self, the life in the fourteenth century progresses alongside the life in the twentieth century. Both exist at once. Development of *each* life continues after physical death, but not along the time axis as we interpret it. There exists a perpetual mutual interplay between the various lives, a relationship that takes place through the higher-dimensional self. The situation becomes even more complex when we realize that the moment point also includes future thoughts and actions. Our present life is not only affected by our past lives but by our future lives as well. The probable experiences of each life alter the experiences of all the others. If we understand that all the reincarnational lives are actually aspects of one entity, then the whole self is changed by all of its comprehensions. Yet paradoxically, each life is unique, and independent, and determines its own destiny even after it leaves our earthly time line.

CONCLUSION

The picture of reality presented by our three sources is a great deal richer than what we see using the lens of our existing paradigm. Both the physicist and the mystic have encountered an underlying multidimensional reality. Quite naturally, each reacts to it in a distinct way. The mystic returns from meditation and proclaims that our views are simplistic, but there is no language adequate for showing us the true way. The physicist uses mathematical concepts to describe the greater reality, but is reluctant to acknowledge the significance of these discoveries. The mathematical structures used so effectively are described as mere algorithms, not to be taken seriously except as tools for manipulation.

David Bohm, however, has studied the algorithms and has seen their significance — and therein lies his genius. He is like the young child who declared that the emperor has no clothes. If we only remove our self-imposed blinders, he implies, untold and unimagined aspects of the unseen reality can be ours to see and experience. Finally, Seth bridges the mathematical concepts of the physicist and the inexpressible experience of the mystic with a detailed vision that focuses the two into a single image.

4

Common Elements

Assuredly, all the disciplines of the mind and all the sciences of man are equally precious and their discoveries mutually so, but this solidarity does not mean confusion. What is important is to integrate the results of the diverse applications of the mind without confounding them. The surest method . . . is still that of studying a phenomenon in its own frame of reference, with freedom afterwards to integrate the results of this procedure in a wider perspective.

Mircea Eliade

As we have seen, the idea of levels of reality is shared by David Bohm, the Perennial Philosophy, and Seth.[1] Nick Herbert postulates an analogous system of three levels, each of which is determined by its accessibility to physical instruments. Level one is the level of the quanta, the domain of real quantum events, our three-dimensional world. Level two is "unreal" in that it is the level of probability, where all choices, including contradictions and opposites, exist in a live-and-let-live manner. This is the place of Heisenberg's *potentia*, the home of the wave function before it collapses. (All possible quantum states are represented together in the wave function, with each state having a given probability of existence in our three-dimensional world.) This is also the location of Einstein's "ghost field." Herbert's third level, which he calls level zero, is where the choice is made among the myriad selections available on level two. The probability that is chosen manifests on level one as a real event and is made by what Herbert calls the choreographer, analogous to Bohm's formative cause or the superimplicate order.

While quantum theory describes level one and assumes level two, physics in general has nothing to say about level zero, which is the level of the hidden variables. Among the physicists who have talked about this level, of course, is David Bohm. Based on his interpretation of the formalisms of quantum theory, Bohm's view of our three-dimensional universe — the explicate order — is that it is not complete. It rests on, or is floating in, an implicate order. The explicate order is equivalent to Herbert's level one, the implicate order to level two, and the superimplicate order to level zero. But Bohm does not stop there. He postulates an unending spectrum of orders that merge into the totality of the holomovement. To Bohm, levels one and two are endlessly repeated throughout the holomovement (at different degrees of subtlety) along with level zero, which determines what Herbert calls the choreography. Each level becomes its own explicate order arising out of its own implicate order and directed by its own superimplicate order.

What is this spectrum? Bohm calls it a sea of cosmic energy. On the lowest level of the spectrum, we have the elementary particle or matter, to which Bohm applies the term "protointelligence." That is, the electron must be at least structured enough to respond to the information generated by the superimplicate order. As the depths of the holomovement are explored, the intelligence factor increases proportionately. Somewhere in the holomovement lies the source for sentient consciousness. If we assign a protointelligence to matter and postulate greater intelligence as matter becomes more subtle, then Bohm's spectrum of orders becomes a spectrum of consciousness.

REALITY AS LEVELS OF CONSCIOUSNESS

The role of consciousness in quantum theory has been explored by several physicists. The fact that the collapse of the wave function is due to the observer has catapulted consciousness into any serious discussion of the meaning of the collapse. Among those physicists who place consciousness in Herbert's level zero is Evan Walker. To him, the hidden variables turn out to be the choices of all human consciousness.[2] Rather than stretching out level zero into a spectrum, as Bohm does, Walker sees level zero as the home of all consciousness.

The essential point is that Bohm, Herbert, and Walker agree that the three-dimensional world needs a nonphysical level to satisfy the formalisms of quantum theory. Bohm says we live in a world of infinite levels, while Walker and Herbert say we live in at least a three-level world. All introduce sentient consciousness, but Bohm places it deeper into the holomovement. Herbert and Walker deal only with the one level of human consciousness rather than the entire spectrum of consciousness. To Bohm it is quite clear: the explicate order is definitely an aspect of the implicate order and the entire holomovement. If the holomovement contains or enfolds consciousness, then in some manner or degree, matter must be an aspect of consciousness also.

As we have seen, the basic feature of the Perennial Philosophy is that consciousness is displayed as a hierarchy of levels. Wilber compares the spectrum of consciousness to the electromagnetic spectrum, in which the entire spectrum from gamma rays to radio waves shares a common set of properties — all are electromagnetic waves, and all travel at the velocity of light — and are differentiated only by their frequencies and energy levels. In a similar manner, the spectrum of consciousness includes numerous levels which differ in degree of subtlety (or density). Wilber recounts this description by Lama Govinda, a Tibetan Buddhist:

> These levels are not separate layers . . . but rather in the nature of mutually penetrating forms of energy, from the finest "all radiating," all pervading, luminous consciousness down to the densest form of "materialized consciousness," which appears before us as our visible, physical body.[3]

Note the characterization of matter as "materialized consciousness."

Wilber's division of the spectrum of consciousness into six levels is, of course, arbitrary. It is important to remember that these levels are not conceived as separate or distinct but mutually interpenetrating and perhaps infinite in number. Bohm also insists upon this in his concept of wholeness: all levels are enfolded in each other. Bell's theorem, too, indicates that everything is interconnected on some subjacent level. The idea of wholeness and interconnectedness, then, is a salient feature in the conceptions of both Bohm and the Perennial Philosophers and is found in physics as well.

THE PARADOX OF LEVELS WITHIN WHOLENESS

The notion of levels within wholeness is difficult to visualize. Perhaps we can say that it is a matter of focus. To Bohm, we are aware of the holomovement only by what is explicated (although he does not rule out the possibility of penetrating the depths of the holomovement by extraordinary states of consciousness). That is, the explicate world is the aspect of the holomovement that we are focused upon. A. P. Shepherd, as quoted by Wilber, states that we are unable to experience the higher levels even though they interpenetrate us. So while all levels are one and wholistic, we explicate or experience a definite aspect, depending on our level of consciousness.

Wilber emphasizes that each higher level transcends and includes the next lower level. Because of this transcendence, the higher plane cannot be derived from the one beneath it. The higher level contains all the aspects of the lower level, but also exhibits properties that are clearly different from the lower one. These added properties demonstrate that we cannot determine the higher level in a reductionist manner.

Wilber shows some irritation with the notion that the spectrum of consciousness can somehow be ascertained through the study of physics. As he frequently states, physics studies the lowest level and no more. Wilber thus denies the possibility that the physicist can come to essentially the same conclusions as the mystic without going through the mystical process. But perhaps Wilber misunderstands what Bohm is attempting to say.

According to quantum theory, Bohm says that space is not a vacuum but a vast ocean of energy. The ocean of energy is not in space-time; it is in the implicate order and is unmanifest. The manifest portion of the ocean of energy is matter, which can be seen as a ripple on the ocean. Bohm then postulates that, based on quantum formalisms, there *may* be still more oceans of energy that enfold the implicate order. To quote from a conversation between Bohm and Renée Weber:

Weber: But the source, you're saying, is in the implicate order and that's this ocean of energy, untapped or unmanifest.

Bohm: That's right. And in fact beyond that ocean may be still a bigger ocean because, after all, our knowledge just simply fades out at that point. It's not to say that there is nothing beyond that.

Weber: Something not characterizable, or unnameable?

Bohm: Eventually, perhaps, you might discover some further source of energy but you may surmise that that would in turn be floating in a still larger source and so on. It is implied that the ultimate source is immeasurable and cannot be captured within our knowledge. That really is what is implied by contemporary physics.[4]

This is not to say that a hierarchy is derived from the laws of physics, but that a study of physics suggests that even though we do not have knowledge of other levels, such levels are not ruled out. Furthermore, Bohm is not reducing mind to matter or matter to mind. That would be a form of reductionism. Rather he sees each as an aspect of the other; they are interwoven. The following exchange between Weber and Bohm clarifies this point.

Bohm: Another way of saying that is that everything material is also mental and everything mental is also material, but there are many more infinitely subtle levels of matter than we are aware of.

Weber: Or that we ever could be aware of?

Bohm: Yes. We could think of the mystic as coming in contact with tremendous depths of the subtlety of matter or of mind, whatever we want to call it.

Weber: A depth where the distinction no longer applies?

Bohm: Yes. Rather than use the word "contact," we can say the mystic enters into it.

Weber: Plato says "to behold the form and enter into union with it."

Bohm: Yes. If you don't distinguish mind and matter then it becomes conceivable that you can enter into it. If you believe that matter is purely material, then how are you going to enter into it?

Or if you think of it as purely mental then we have to think of it as some far away thing and some mysterious leap is required. But what if one supposes that actually it's not far away at all.[5]

Wilber subdivides each level into a deep structure and a surface structure. He defines the deep structure as containing all the potentials along with the limitations of that particular level. The surface structure of the level is a distinct manifestation of the deep structure. The surface structure can unfold in any form or use any potential within the limits of that specific deep structure. The deep structure contains the rules of the game; the surface structure is the display. When these general concepts are applied to the material level, we find we are describing the explicate and implicate orders. To quote Wilber:

> In short, the implicate order as I would state it, is the unitary deep structure of level 1 [matter, energy, etc.] which subscends or underlies the explicate structures of elementary particles and waves.[6]

The question is, do Bohm's views have application beyond the lowest level of the Perennial Philosophy? Recalling that the deep structures of all levels are not separate but interpenetrated and enfolded, we see that they are aspects of the same whole. Wilber identifies this as the ground unconscious. He says:

> All the deep structures given to a collective humanity — pertaining to every level of consciousness from the body to mind to soul to spirit, gross, subtle and causal — are enfolded or enwrapped in the ground unconscious.[7]

The ground unconscious, then, is analogous to Bohm's holomovement. The holomovement enfolds all the levels of the implicate orders. While it is true that Herbert's three levels (one, two, and zero) can be identified with Wilber's material level only, it is quite obvious that Bohm's holomovement is more encompassing and applies beyond the lowest level of the Perennial Philosophy.

THE RELATIONSHIP BETWEEN LEVELS

As we noted earlier, Wilber is firm in stating that each higher level transcends the next lower level: "What is the *whole* of one level becomes merely a *part* of the higher-order whole of the next level."[8]

How does Bohm view the relationship between levels? Again, the following is from an interview with Weber:

Weber: How do you order these various levels?

Bohm: To say that the higher level simply transcends the lower level altogether. It's immensely greater and has an entirely different set of relationships out of which the lower level is obtained as a very small part, in an abstraction.

Weber: It has wholeness, more power, more energy, more insight?

Bohm: Yes, and it contains the lower level in some sense.

Weber: And not vice-versa?

Bohm: The lower level will be the unfoldment of the higher level.

Weber: In space and time.

Bohm: Yes.

Weber: So in a sense they contain each other but in another sense the higher one contains the whole and the lower one is more linear.

Bohm: Yes. The higher one is called non-linear, mathematically, and the lower one is linear. That means of course that the linear organization of time and thought characteristic of the ordinary level will not necessarily be characteristic of the higher level. Therefore what is beyond time may have an order of its own, not the same as the simple linear order of time.[9]

One of the basic tenets of the holomovement is the quality of unfolding and enfolding. Through the process of unfolding, the explicate order is created. Through the process of enfolding, the explicate order sends its information to the implicate order. Bohm calls this projection (unfolding) and injection (enfolding). According to his causal theory, the fluctuations are extremely rapid.

For every moment that is projected out into the explicate there would be another movement in which that moment would be injected . . . back into the implicate order.[10]

Bohm sees the universe as populated with a tremendous diversity and number of forms. The forms are *not* matter, but are the shape and organization of matter, the patterns of matter, created and maintained by innumerable repetitions of unfolding and enfolding, until a constant component or shape takes place. It is these forms that are involved in the transformations we observe in our explicate world.

As to how these particular forms are created, Bohm postulates the superimplicate order, a second level. The first level is the implicate order, or the ocean of energy that is capable of creating structure in the explicate order or causing ripples (matter) on the sea of energy. On the next level of enfoldment, the superimplicate order organizes the implicate order into structures. The superimplicate order is nonlinear and is capable of organizing the linear implicate order. The implicate order does not have the capacity to unfold itself.

Again, Wilber is more general regarding levels beyond the material.

> As evolution proceeds . . . each level in turn is differentiated *from* the self, or "peeled off" so to speak. . . . To *identify* with the next higher order . . . . It *transcends* that structure . . . and can thus *operate* on that lower structure . . . .[11]

In Bohm's and Wilber's view of the material level, the next higher plane can operate on the lower one. If the lower level is the material universe and its implicate order, the self using the next higher level (the superimplicate order) can operate on the material world. What Bohm calls the superimplicate order, Wilber calls the self when it occupies a position just above the level it is manipulating. In Wilber's hierarchy, the biological level operates on the next level down, the material level. For an electron, the biological level would be the superquantum potential; for a living cell, the analogous superquantum potential would be the vital principle (*prana* in yoga) of the cell; for the human being, the superquantum potential would be the bodyego (on the mental level; see p. 103). For example, the bodyego would operate on the muscles of the body to help it walk.

MEANING AS FORMATIVE CAUSE

To understand how to operate, as Wilber says, or to organize or unfold, as Bohm terms it, we need more detailed information. Bohm explains this through his concept of soma-significance. Recall that soma (material) and significance (mental) are aspects of *one* overall reality, not two different entities. On the particle level, the particle and its field are the soma and significance of reality. The field has a form that directs the particle to act in a certain way or engage in a particular activity. This activity is the information that the field carries. Bohm calls it "meaning."

It is meaning that guides the electron through its worldly journeys. Meaning is not soma and not significance — it is the bridge between them, the activity of information. Meaning is the score of the dance that the choreographer, in Herbert's terms, uses to induce the dancers to move in some predetermined way. In the case of an electron and its field, meaning is the information used by the superimplicate order to organize the particles. The same approach can be applied to cells or to human beings. In a manner similar to the arrangement of multilevels of implicate orders, meaning also can be viewed as a spectrum in the holomovement. At each level of the holomovement, there is a concurrent level of meaning. Therefore, all the meanings can be enfolded in each other and in orders of infinite extension.

We can better compare Bohm's and Wilber's spectrums if we first examine Bohm's concept of meaning. Bohm defined meaning as the essential feature of conscious awareness. For Bohm, meaning is active and the activity of consciousness is determined by meaning. In these conversations published in *Unfolding Meaning*, the similarity of Bohm's view and that of the Perennial Philosophy becomes more apparent.

> *Bohm:* You could say that consciousness, both in the features that we experience and in its activity, is meaning. And the greater the development of meaning, the greater the consciousness.[12]

At another point, Bohm identifies meaning with being:

> *Questioner:* Do you see our function being to reveal meaning?

> *Bohm:* Well, we reveal meaning in a way which the rest of matter does not. But we are meanings. You see, I think that if you say "reveal meaning," it suggests that meaning is something different from being.

Since Bohm equates meaning and consciousness, he sees the evolution of consciousness as the unfolding of meaning.

> *Questioner:* Would it be right to say that you would define evolution as the movement of the universe towards its own coherence, in terms of what it is or can be?
>
> *Bohm:* Yes. Evolution is the unfoldment of these meanings.
>
> *Questioner:* In terms of greater and greater coherence?
>
> *Bohm:* Yes.
>
> *Questioner:* And subtlety?
>
> *Bohm:* Subtlety, yes.

In a sense, then, meaning climbs the ladder of consciousness. Bohm states it as follows:

> There is a perception of meaning going from one level to another, and a signa-somatic activity backward. But meaning is this whole activity in which the meaning of the soma comes into the next level of subtlety, and action goes back out. Those are some of the essential aspects of meaning — that meaning pervades being. I think that you cannot get the whole meaning of meaning because our ideas of meaning are never fully defined.

Note that Bohm says that meaning (consciousness) goes up to the next level of subtlety, and action goes back down through a signa-somatic process, or the higher mental level acting on the lower physical level. We have, through this discussion, established a more complete concept of meaning, which can now be identified with a form of consciousness and whose subtlety depends on the level. Evolution is the process of unfolding meaning or consciousness through the spectrum. When meaning is unfolded at a given level, it can act on a lower soma level. Bohm sees meaning as a formative cause.

> Meaning operates in a human being as a formative cause; it provides an end toward which he is moving; it permeates his attention and gives form to his activities so as to tend to realize that end.[13]

In another sense, Bohm views the wave function of a particle as a formative cause. He says:

Let us interpret the wave function differently as representing a description of the formative cause, but as having some sort of meaning which would also tie in with things like the non-local correlation. There would be a meaning connected with the whole system as well as with any one part.[14]

So the wave function is a formative cause that has meaning (consciousness) that is supplied by the super-wave-function. The meaning (consciousness) associated with a particle is similar to the meaning (consciousness) of a human being; they differ only in their positions on the spectrum. To Bohm, it is a matter of focus.

If you are primarily focused in the explicate order, your sense of space will be confined to the spaces between a lot of separate objects. As you go further into the explicate order, you begin to see that these objects contain each other and fall into each other. Eventually you will see them as forms within a much vaster space, and finally, a space in which no forms are created. I think this corresponds to different stages of consciousness.[15]

On the particle level, meaning guides the particle through its trajectories. With our more developed definition of meaning, we see that the electron, because of its level on the spectrum of meaning, has a form of consciousness guiding its journeys. To be sure, this is not to be equated with human consciousness, but consciousness it is, nonetheless — a kind of protointelligence. Nature is alive with meaning at all levels. Seth concurs:

There is no such thing, in your terms, as nonliving matter. There is simply a point that you recognize as having the characteristics that you have ascribed to life, or living conditions — a point that meets the requirements that you have arbitrarily set.

This makes it highly difficult in a discussion, however, for there is no particular point at which life was inserted into nonliving matter. There is no point at which consciousness emerged. Consciousness is within the tiniest particle, whatever its life conditions seem to be, or however it might seem to lack those conditions you call living.[16]

As evolution proceeds, we encounter more subtle levels that eventually are equated with human consciousness. Perhaps there are even more subtle levels beyond that.

THE ORIGIN OF FORMS

There is one important difference between Wilber's conception of the spectrum and Bohm's. Wilber states that symbols are potentials in the ground unconscious waiting to emerge, waiting for the self to remember them. This view coincides with that of Plato and mathematician Kurt Gödel, that mathematical formalisms are discovered; they exist in potential form, waiting to be unfolded. As Rudy Rucker says of Gödel's approach:

> There is one idea truly central to Gödel's thought . . . . This is the philosophy underlying Gödel's credo, "I do objective mathematics." By this, Gödel meant that mathematical entities exist independently of the activities of mathematicians, in much the same way that the stars would be there even if there were no astronomers to look at them. For Gödel, mathematics, even the mathematics of the infinite, was an essentially empirical science.
>
> According to this standpoint, which mathematicians call *Platonism*, we do not *create* the mental objects we talk about. Instead, we *find* them, on some higher plane that the mind sees into, by a process not unlike sense perception.[17]

Bohm, on the other hand, clearly holds a different view on this point. Asked by Weber if forms in the implicate order pre-exist in the Platonic sense, Bohm replied that he sees the forms as developing "through the process of projection and injection, re-projection, re-injection, and so on."

Weber: So there is a developmental process in the implicate order?

Bohm: Yes. A development of form.

Weber: Where does creativity come into this notion of projection and injection and re-projection?

Bohm: As far as the implicate order is concerned, every new moment could, in principle, be entirely unrelated to the previous one — it could be totally creative. You could say that creativity is fundamental in the implicate order, and what we really have to explain are the processes that are *not* creative. You see, usually we believe that in life the rule is uncreativity, and occasionally a little burst of creativity comes in that requires explanation. But the implicate order turns all that around and says creativity is the basis and it is repetition that has to be explained. That's where the morphogenetic field theory comes in.[18]

Biologist Rupert Sheldrake's morphogenetic field theory provides a specific example of the implicate order concept and its property of projection and injection. In 1981, Sheldrake suggested in his book *A New Science of Life* that behind the myriad forms encountered in biology, there are morphogenetic fields that create and control the numerous gestalts of living organisms. These fields are nonenergetic and nonlocal, which means that they are not part of our three-dimensional world.

The parallel between morphogenetic fields and quantum fields of elementary particles is apparent. Recall that the de Broglie waves are associated with all forms of matter. Just as each particle has its own quantum field, each form has its own morphogenetic field. Through a process of projection and injection (unfoldment and enfoldment), the organism created by the field also influences the field. Thus morphogenetic fields are not fixed; they are subject to change through this feedback system. The fields are involved in an evolutionary development. The form may be considered as existing prior to the organism, but it is influenced by what it creates. Sheldrake feels that the origination of form is beyond the present purview of science, since biology does not attempt a description of the implicate order.

Bohm stresses the creativity of the holomovement and our part in it. In essence, the implicate order is not a storehouse of potentials waiting to be discovered; it is a constantly changing spectrum of orders. The explicate order plays a key role in this change by displaying the implicate order and thus creating an injection of information from the explicate order to the implicate order. In this case, Seth agrees with Bohm. He comments:

> You imprint the universe with your own significance, and using that as a focus you draw from it, or attract those events that fit your unique purposes and needs. In doing so, to some extent you multiply the creative possibilities of the universe, forming from it a personal reality that would otherwise be absent, in those terms; and in

so doing you also add in an immeasurable fashion to the reality of
all other consciousness by increasing the bank of reality from which
all consciousness draws.[19]

Bohm might sum up Seth's statement by saying that unfoldment enriches what is enfolded.

GÖDEL'S INCOMPLETENESS THEOREM AND INFINITY

Both Bohm and Wilber postulate a hierarchical spectrum as a
metaphor for reality. Perhaps the logic of this construct can be seen
from the notion of incompleteness first put forth by Gödel. In 1931, at
the age of twenty-five, Gödel published a paper in logic called "The
Incompleteness Theorem." The full implications of this theorem have
still not been exhausted. While the original paper applied to mathematical logic, the theorem applies to all formal systems of knowledge. Basically, the theorem concludes that on the deepest level, reality is infinite. One definition of this theorem is that

> Within any rigidly logical mathematical system there are proposi-
> tions (or questions) that cannot be proved or disproved on the basis
> of the axioms within that system and that, therefore, it is uncertain
> that the basic axioms of arithmetic will not give rise to contradic-
> tions.[20]

In simpler terms, Gödel's theorem states that when mathemati-
cians formulate propositions and develop a system of mathematics,
the truth or falsity of these propositions cannot be ascertained by the
mathematics of that system. Beyond a certain level of complexity (and
this need not be highly complex), there are intrinsic limits to the truths
that can be established within the system. If, however, we can view
the system from the outside, or within a wider context, the original
system's statements can be provable. But in so doing, the wider context is now subject to the incompleteness theorem and again has to be
viewed from an even wider context. This, of course, defines a spectrum of contexts in which each level transcends the context below it.
There does not seem to be a way to deductively reason our way up the
ladder of the spectrum, and indications are that the spectrum would
have to be infinite.

Gödel's theorem can also be understood from a self-referencing point of view. If anything is to see itself, it must literally cut itself in two. One part becomes the object and the other part the viewer. The portion called the viewer is not part of the object to be viewed and as a result is not seen. The total self is then incompletely described. To overcome this problem, the total entity must be seen "outside" itself or from a wider context. In essence, the original total self must be transcended. As one can see, this method is never-ending; the requirement for transcendence goes on indefinitely because a portion is always missing. Knowledge is never complete, so we end up with an infinite hierarchy of levels. As the French artist and writer Jean Cocteau phrased it, "There are mysteries within mystery, gods above gods . . . . That's what is called infinity."[21]

A paradox arises if matter is seen from the point of view of materialism. When the universe is thought of as made up of objects, it is so viewed by the mind. If the brain is an object, and the mind is an attribute of that object, the mind ends up observing itself. Again, the viewer cannot be included in the universe that is viewed. To resolve this paradox, we conclude that brains and minds must be epiphenomena of something else. This again starts the process of creating a spectrum. The whole self-referencing paradox generates the necessity for an order beyond our three-dimensional world.

If Gödel's theorem is correct, a system of knowledge like quantum theory (or any theory) must be, by definition, incomplete. The only way to extend our knowledge indefinitely is to use a philosophical backdrop such as the spectrum of consciousness or implicate orders. We are, according to Gödel, compelled to move through the spectrum if we are to create ever more encompassing systems of knowledge. Both Wilber and Bohm have come to the same conclusion.

TIME AS A CONSTRUCT

Bohm concludes that both space and time are derived from a higher-dimensional ground, the holomovement. Time is to be viewed as a particular type of order that is explicated from its own implicate order. The ground cannot be comprehended in any particular time order. Since time is explicated in a way similar to matter, any given moment of time has the whole of time enfolded within it. Both matter and time exhibit different levels of explication, with the connections (forms and moments) not in space but in the implicate order. Time, then, becomes a succession, or an aspect of succession, in the depth of implication. As you go into the holomovement, the events that will appear successively are actually co-present. In short, time is a metaphor for the depth of implication rather than a steady flowing stream inherent in the explicate order.

Before reviewing Wilber's approach to time, we might reiterate the fact that classical physics was based on the assumption that space had three dimensions, all independent of the material in it. Time was viewed as a separate dimension. Furthermore, time was seen as flowing at a constant rate and was considered independent of the material constituting our world. These aspects of time were not thought of as mental concepts, but were seen as basic features of reality itself.

In 1905, Einstein completely revised the classical notions of time with his special theory of relativity. For one thing, he formally connected space and time into one entity, space-time. Henceforth, all discussions about space involved discussions about time, and vice-versa. Einstein also abolished the concept of a universal, linear flow of time, since different observers may perceive different sequences for events when they move with different velocities relative to the observed events. The important point for our discussion is that space and time are constructs of the mind of the observer rather than inherent in any absolute way in the fabric of reality.

Wilber and the Perennial Philosophy see time as a product of the third or mental level in a similar way. As mentioned previously, language is one of the symbolic structures of this level. It is through language that sequential arrangement is possible. To quote Wilber:

> The development of language — the symbolic structures of word-and-name — brings with it the ability to recognize series of events and sequences of actions, and thus to perceive the salient non-present world. In other words, the symbolic structures of language *transform* the present moment into a *tensed* moment, a moment surrounded by the past and the future. Thus does word-and-name transform the passing present of the axial-body level into the tensed duration of the verbal-membership level. It allows consciousness to

transcend the present moment, which is a decisive and far-reaching ascent. And — to cap this brief discussion — the next major symbolic structure, the syntaxical thought, creates a clear and enduring mental structure of times past and times future. So it is that at each stage of evolution an appropriate symbolic structure — itself emerging at that stage — transforms each particular mode of time into its successor, and thus marks the pace of the ascent of consciousness.[22]

Wilber agrees with relativity physics that time is a mental fabrication, and he agrees with Bohm that time becomes a succession of moments, in that the self recognizes a "series of events" and "sequences of action." Bohm sees all of time enfolded in each moment of time; Wilber says that "all of eternity is completely present at every point of time" and that "the only Reality is present Reality."[23] Evidently both agree that the past and future are contained in the present.

THE CREATION OF MATTER

Bohm was led to the process of projection and injection by reformulating present-day quantum mechanics. If we think of space as an ocean of energy, a wave arises or is projected from the ocean and then recedes or is injected back into the ocean. For every moment projected out into the explicate order, there is a moment injected back into the implicate order. If this process is repeated a large number of times, a fairly constant wave pattern is formed in the explicate order. This is what we term matter. Each cycle reinforces the pattern, but it can change slightly through the nonlocal connections within the implicate order. Not only is each new cycle influenced by past cycles, but also the pattern can change by the creativity inherent in the implicate order. By creativity, we here mean the choice or thought operating at Bohm's superimplicate level (Herbert's choreographer). When a stable pattern is not developed (not enough cycles of projection and injection), the process allows for this creativity. If the cycling is relatively stable, a stable form of matter is created. This process also allows for the observed continuity of the explicated world.

Thus, if all projections were only on the creative level of the superimplicate order, everything would disappear as soon as it appeared. There would be no past. On the other hand, without creativity, we would be mired down in our past and become machinelike. It is the combination of both processes that allows the universe to experiment and learn.

If this concept of projection and injection is applied on the particle level, the quantum field is seen in a state of very rapid fluctuation. Bohm says that "the fluctuations of the psi-field [implicate order] can be regarded as coming from a deeper sub-quantum mechanical level . . . ."[24] He compares the field's oscillations to that of the Brownian motion of a microscopic liquid droplet which comes from the atomic level. Since the particle is actually a pattern of fluctuations in the quantum field, it literally projects outward into our three-dimensional world and inward to a subquantum level. Thus, we see a particle as a pattern of projections and injections of the implicate order. A whole array of particles, such as atoms and molecules, will behave in a similar manner. Matter is actually a rapid succession of almost instantaneous events, blinking in and out of the universe at an incredible rate.

Wilber points out that according to the Perennial Philosophy there are two major movements: evolution and involution. He describes evolution as the climb of the self through the various stages of the spectrum of consciousness. The goal of the self is to ascend from the lowest or material level upward to ultimate unity. Involution is exactly the opposite. In this process, the ultimate separates itself, or throws itself outward, to create the various manifest worlds. Involution is how the higher orders of consciousness come to be enfolded in the lower states.

This whole process of evolution and involution goes on not only through the ages but moment by moment. The higher levels of the spectrum actually generate the lower levels, including the manifest world, by intermittent projections of energy. The time between projections is rather vaguely given as the "twinkling of an eye." Similarly, Seth speaks of pulsations, or continuous creations of psychic energy into physical patterns.

THE LEVELS CORRELATED WITH SETH'S FRAMEWORKS

Seth's views reinforce those of both Bohm and Wilber in that he also describes a hierarchy of realities. The idea of levels of reality was introduced into physics in the twentieth century through the EPR thought experiment (described in Chapter 8) and quantum theory. But the straw that broke the camel's back of classical physics in this regard was Bell's theorem. The nonlocality requirement made the postulation of a level beyond space-time both necessary and acceptable. It forces us to face the fact that space-time is not a closed system; a second level, at least, is in some way involved with our three-dimensional world. This theme runs throughout Seth's books. He says:

> Basically . . . no system is closed. Energy flows freely from one to another, or rather permeates each. It is only the camouflage structure that gives the impression of closed systems, and the law of inertia does not apply. It appears to be a reality only within your own framework and because of your limited focus.[25]

That is, the universe of Newtonian space-time is *not a closed system*. It is an integral part of a wide range of realities that are interdependent and essentially merged.

Seth arbitrarily designates the various levels as frameworks but cautions against too much reliance on any demarcations since the levels are utterly interpenetrated. His construct clearly is similar to the spectrum of implicate orders and the hierarchy of consciousness. Although he discusses only four frameworks, Seth clearly indicates that this enumeration can extend without limits. "Behind [Framework 2]" he says, "are endless patterns of orderliness and complexity that are beyond your conscious Framework 1 comprehension."[26]

An obvious correlation exists between Bohm's explicate order, Wilber's physical (material) level, and Seth's Framework 1. This level stretches from electrons to stars, encompassing the world in which we carry on our daily lives, the everyday world of experience, the world of display. Seth's Framework 2 can be identified with Wilber's deep structure and with Bohm's implicate order. According to Seth, Framework 2 houses the probabilities that are available to Framework 1. Thus all the events of Framework 1 emerge from Framework 2, which is the creative medium responsible for physical life. In Framework 2, the moment point is operative, as it is in the implicate order and in the deep structure.

In Framework 3 we find Bohm's superimplicate order, Wilber's higher level operating on the lower one, and the level zero of Herbert's choreographer, who chooses among the possibilities available on level two. Every electron, every cell, every animal, every human being has a choreographer in Framework 3, the abode of the outer ego. It is in Framework 3 that Jane Roberts and Seth actually came together. Seth says:

> The encounters themselves occur in a Framework 3 environment. That framework of course, again in terms of an analogy, exists another step away from your own Framework 2. I do not want to get into a higher-or-lower hierarchy here, but the frameworks represent spheres of action. Our encounters initially take place, then, beyond the sphere that deals exclusively with either your physical world or the inner mental and psychic realm from which your present experience springs.[27]

Framework 4 involves a still broader focus and corresponds to Bohm's super-superimplicate order. Perhaps it is also equivalent to Wilber's subtle level. Seth describes it as follows:

> Framework 4 . . . is somewhat like Framework 2, except it is a creative source for other kinds of realities not physically oriented at all and outside of, say, time concepts as you are used to thinking of them. In a way impossible to describe verbally, some portion of each identity also resides in Framework 4, and in all other frameworks.

> Some invisible particles can be in more places than one, at once. Some portions of each identity can also be in more than one place at once. It is a matter of focus and organization.[28]

In regard to these higher frameworks, Seth comments:

> This is what I know of reality. There is far more to be known. Outside of the realities of which I am aware and others are aware, there are systems that we cannot describe. They are massive energy sources, cosmic energy banks, that make possible the whole reality of probabilities.

> They have evolved beyond all probabilities as we understand them yet, outside of probabilities, they still have existence. This cannot be explained in words. Yet, none of this is meant to deny the individual, for it is the individual upon whom all else rests, and it is

from the basis of the individual that all entities have their existence. Nor are the memories or emotions of an individual ever taken
from him. They are always at his disposal.[29]

Using terminology that recalls Bohm's spectrum of orders, Seth says:

> True spontaneity . . . comes directly from Framework 2, and behind
> it are endless patterns of orderliness and complexity that are be-
> yond your conscious Framework 1 comprehension . . .[30]

The correlation of Seth's Framework 2 with Wilber's deep structure can be more readily understood by the following quotation from
Wilber in which he defines the relationship between the deep structure and the surface structure. Keep in mind that the surface structure is analogous to Framework 1.

> Modifying the terms of linguistics, we can say that each level of
> consciousness consists of a *deep structure* and a *surface structure*.
> The deep structure consists of all the basic limiting principles em-
> bedded at that level. The deep structure is the defining form of a
> level, which embodies all of the potentials and limitations of that
> level. Surface structure is simply a *particular* manifestation of the
> deep structure. The surface structure is constrained by the form of
> the deep structure, but within that form it is free to select various
> contents (e.g., within the form of the physical body, one may select
> to walk, run, play baseball, etc. What all of those forms have in
> common is the deep structure of the human body).
>
> A deep structure is like a paradigm, and contains within it all the
> basic limiting principles in terms of which all surface structures
> are realized.[31]

Note that Wilber says the deep structure contains the "limiting
principles" or constraints for the surface structure. Seth talks about
the operation of free will on the level of our Framework 1 (Wilber's
surface structure, Bohm's explicate order). Seth observes:

> This is, again, difficult to explain, but free will operates in all units
> of consiousness, regardless of their degree — but it operates within
> the *framework* of that degree. Man possesses free will, but that free
> will operates only within man's degree — that is, his free will is
> somewhat contained by the frameworks of time and space.

He has free will to make any decisions that he is *able to make*. This means that his free will is contained, given meaning, focused, and framed by his neurological structure. He can only move, and he can only choose therefore to move, physically speaking, in certain directions in space and time. That time reference, however, *gives* his free will meaning and a context in which to operate. We are speaking now of conscious decisions as you think of them.[32]

Framework 3 might be considered the region entered by a person's consciousness upon death. In the literature of the near-death experience, a breaking away of a second body from the material one is often described. This occurrence has been labeled the out-of-body phenomenon. In accounts of the near-death experience, many people talk about traveling through a tunnel or dark void. Itzhak Bentov, author of *Stalking The Wild Pendulum*, talks about the "adjustment of the consciousness from one plane of reality to another."[33] Seth calls it a sort of refocusing:

> In your terms, the dead waken in Framework 2 and move through it to Framework 3, where they can be aware of their reincarnational identities and connections with time, while being apart from a concentration upon earth realities.[34]

PARTICLES AS CONSCIOUSNESS UNITS

Seth says that reality is formed from elementary consciousness units (CUs), the arrangement and form of which result in our universe. This seemingly reductionist approach falls apart when the CU is defined. According to Seth, CUs have definite propensities and purpose; therefore, the basic building blocks of the universe are *conscious*. In order to avoid confusion, we must remember that the consciousness of the CU is not like that of a sentient being. The CU is not inert, however, and it has the uncanny ability to appear as a particle in our three-dimensional universe and to swarm as a wave outside our universe.

The similarity between Seth's concept of CUs and Bohm's view of particles is apparent. To Bohm, the elementary particle is so constituted as to enable it to react to active information found in the wave function through its quantum potential. Basil Hiley, Bohm's collaborator, makes clear that this does not mean that the elementary particle has cogs and gears on a microscopic level; this would be returning to the Newtonian model of inert billiard balls. Rather, the elementary particle should be viewed as an aspect of the whole as defined by quantum field theory. The particle is a manifestation of the field (or implicate order) in our three-dimensional space and as such it has a kind of protointelligence. This intelligence is evident because the implicate order is itself alive.

To Seth, the particle is a particular gestalt of CUs as they appear in our world, following definite methods and laws. In the Perennial Philosophy, matter is a projection of the underlying ground. Matter is the same "stuff" as the deep structure, but because it is at the lowest rung on the consciousness ladder, it is labeled by Wilber "insentient slumber." (Recall Lama Govinda's term, materialized consciousness.)

In summation, reality is consciousness that manifests itself in myriad gestalts of CUs, each displaying its own level of awareness. Seth notes the identity of reality and consciousness clearly in the following passage:

> I can tell you . . . that there is consciousness even within a nail, but few of my readers will take me seriously enough to stop in midsentence, and say good morning or good afternoon to the nearest nail they can find, stuck in a piece of wood.
>
> Nevertheless, the atoms and molecules within the nail do possess their own kind of consciousness. The atoms and molecules that make up the pages of this book are also, within their own level, aware. Nothing exists — neither rock, mineral, plant, animal, or air — that is not filled with consciousness of its own kind. So you stand amid a constant vital commotion, a gestalt of aware energy, and you are yourselves physically composed of conscious cells that carry within themselves the realization of their own identity, that cooperate *willingly* to form the corporeal structure that is your physical body.
>
> I am saying, of course, that there is no such thing as dead matter. There is no object that was not formed by consciousness, and each consciousness, regardless of its degree, rejoices in sensation and creativity. You cannot understand what you are unless you understand such matters.[35]

Furthermore, the extraordinary ability of the CUs to unite and swarm creates an unending spectrum of consciousness. Bohm was led to an infinite spectrum of orders by extending quantum field theory. The Perennial Philosophers postulate an endless spectrum wherein the self climbs the ladder, transforming itself at each level and engulfing the ladder as the climb proceeds, a process achieved through meditation or altered states of consciousness. Seth defines the spectrum as a series of frameworks that go beyond anything we or even Seth himself can imagine or understand. Yet with all this, the spectrum does not literally exist. It is a construct that is useful in helping us see that consciousness in all forms seems to find a place in relation to all other consciousness.

The notion of force in Newtonian mechanics is transformed by Bohm's concept of active information. The universe is no longer seen as being made of inert particles that are pushed and pulled by unseen forces. Rather, according to Bohm, information fields influence the manifested particles and guide them through their journey. The analogy that comes to mind is computer hardware, but with a twist. Here the hardware is alive, manifested in our three-dimensional world. The hardware responds because it is conscious; hence it does not need to be as complicated as a conventional computer. In a sense, even a rock is a computer since it is made up of manifested CUs that, in that particular gestalt, respond to the active information that slowly causes it to erode and change over the centuries. The rock has its own implicate and superimplicate orders and occupies its own unique position on the consciousness spectrum.

Einstein recognized this aspect of freedom in quantum theory, but could not reconcile it with his philosophical outlook. In fact, he was quite disturbed by it. Referring to Bohr's Copenhagen interpretation, Einstein wrote the following to Max Born:

> Bohr's opinion about radiation interests me very much. But I should not want to be forced into abandoning strict causality without defending it more strongly than I have so far. I find the idea quite intolerable that an electron exposed to radiation should choose *of its own free will*, not only its moment to jump off, but also its direction. In that case, I would rather be a cobbler, or even an employee in a gaming-house, than a physicist.[36]

In physics, there is no need to define human consciousness in great detail. The superimplicate order is all that is necessary, and perhaps a mathematical expression of it will eventually be worked out.

MOTION AS SEQUENTIAL PROJECTIONS

A new way of looking at motion follows from our new concept of space as a vast ocean of unmanifest energy, identified with the implicate order. Space may appear empty and smooth from the perspective of current measuring instruments, but if we could see down to the subquantum level of the Planck length, Wheeler and other physicists suggest that we would find turbulent activity, with projections and injections of energy from the underlying source occurring at astounding rates. Matter, in these terms, is constituted of energy projections that have reached the point of relevant closure, which we can see with our eyes and measure with instruments.

Motion, then, is understood not as the movement of a discrete unit of matter through empty space but as a sequence of energy projections that changes position within the ocean of energy — in the same way that a ripple appears to move down a river, while what moves is only the form created by water molecules projecting up and down at successive points.

Understanding motion as a sequence of projections gives us a new way of considering interaction. Just as the interaction of two particles is actually the coming together and coming apart of two projections from the implicate order and is governed by the superimplicate order of the ensemble of the two particles, so is the interaction of two people governed by the combination of the two egos. But even that is oversimplified. Since all are projections, and the interaction is controlled by a level above the implicate order, two particles, or two people, cannot, in fact, be separated. All interactions are dependent on all other interactions, since every part of the implicate order contains the whole. So the movement of separate projections is really an aspect of a total movement of the whole spectrum.

Bell's theorem and quantum theory point to undivided wholeness. So do the Perennial Philosophy and Seth. We are all aspects of one undivided whole. In that sense, "nothing" is a meaningless concept. Everything is part of the holomovement, also called All That Is.

CONCLUSION

The idea of probable worlds at first seems rather bizarre. But the superposition of probabilities in the wave function from quantum theory makes this idea at least imaginable. We now know that each event contains the seeds of innumerable probable outcomes. By a selection process, one event becomes actualized and the others are discarded. Everett and Seth indicate that the other events are not lost but go on to form other universes. Bohm assumes that the other probabilities are randomized and therefore rendered impotent, at least for our universe. Everett grants the reality of the probabilities not selected, but suggests that they go their own way and do not affect us. Seth disagrees. He says the probable events not selected do indeed create other universes, and also become part and parcel of Framework 2 and so affect us. The path not chosen is not active in our universe but is there and is available to the choreographer. In any case, all the other realities dip into a common Framework 2, and what is not used here most likely is used elsewhere. In some fashion, the holomovement experiences all probabilities.

Wilber's concept of microgeny fits in with the idea that the universe is a ripple on the vast ocean of the implicate order. At a frequency that we literally cannot track, each particle of matter and all the gestalts that they form project out into the universe and inject into the implicate order moment by moment. All three of our sources, Bohm, Wilber, and Seth, confirm this operation. It is by this method that information and energy are fed into the three-dimensional universe and then injected into the implicate order. Our universe is definitely not a closed system. Energy is constantly being replenished and renewed, moment by moment. Wilber says that each instant the injection goes all the way up the spectrum and returns so that all levels are fed by the process. In this way, all information is available everywhere all the time.

With this view, aging takes on a new meaning. There is no aging of matter as we now see it; rather there is a wearing away or change in the form over time. The actual content of the form exists only for an extremely short moment and then is replaced in the next projection. The apparent persistence of matter is really a persistence of form. So in a sense we, and an electron, die moment by moment. When our form is destroyed, we die (in earthly terms); likewise, the electron vanishes. But since the form is a product of the implicate order, it dies only in the everyday world where projections define it. The form continues to have meaning in the implicate order. Seth says that all gestalts, once formed, never disappear, except from Framework 1. The same system that projects and injects matter into our universe also operates on the level of the mind. Our ego goes through the same process. When in the same way our body flickers in and out of the explicate order, the choreographer goes in and out with it. Our form may, at some point, no longer exist in this world, but, according to Seth, it continues its existence elsewhere.

The concept of time flows right out of the process of projection and injection. The implicate order contains all events at once. All events are interconnected instantly; the need for time is eliminated. When events are selected for a reality such as ours, a sequence of selection defines time. Time is the succession of projections. When the wave function contains all the probabilities, time becomes ambiguous; that is, the wave function can be run equally well forward or backward in time. But as soon as a selection is made and the wave function is collapsed, the symmetry of time is broken and an irreversible process takes place. Time now has a direction and comes into existence. Consciousness produces time by making a selection. As Seth says:

> What separates events is not time, but your perception. You perceive events "one at a time." Time as it appears to you is, instead, a psychic organization of experience. The seeming beginning and end of an event; the seeming birth and death, are simply other dimensions of experience as, for example, height, width, depth.[37]

Time is simply another dimension of a multidimensional reality.

Wilber assigns time to the sentient level because the human mind has the capability to sequence events. Actually, time is created on all levels of the three-dimensional world merely by the creation of the universe through projection. The wave function, which is time-symmetrical, is immune to the flow of time since, in essence, it is outside our universe. As soon as it becomes manifest as matter, time is created. In a sense, matter creates space and time. By its projection, the space in between is defined, and its succession leads to the concept of time. If the ocean of energy were perfectly smooth, we would not be aware of space or time. There would not be a universe. All would be in the timeless implicate order.

We have completed our comparison of the views of David Bohm, Ken Wilber, and Seth, and have uncovered some common elements. In the final chapter of the book, we will look more closely at these common themes and construct a tentative paradigm for reality. In the meantime, in Part Two we will examine in more detail the possibility of reality being created by explication from an underlying order. But, as we emphasize throughout this book, truth is infinite, and our understanding will always be finite.

PART TWO

Selected Topics in Physics and Philosophy— A New Focus

5

Space-Time Creation
and the Black/White
Hole Metaphor

---

*A*s we have seen in Part One, David Bohm believes that our familiar three-dimensional universe is a manifestation of an underlying implicate order and is maintained by a constant process of unfolding and enfolding of events in a vast sea of energy. When a pattern develops in this energy, matter is formed. In a similar manner, Wilber uses the term translations to denote the development of the surface structure from the deep structure. Pulsations of matter that sustain our reality are described by Seth in terms that are comparable to Bohm's idea of unfolding and enfolding. Thus, from the central concepts of these three views there begins to emerge a single coherent metaphor. In Part Two, we examine some of the most challenging issues in contemporary physics and philosophy in light of this new paradigm.

Seth uses black holes and white holes as a metaphor to discuss the pulsing of matter in and out of our universe. In this chapter, we will examine in a qualitative way some of the physical concepts required to determine if Seth's metaphor corresponds to the black/white hole as envisioned by some members of the physics community. Our approach will be to uncover similarities between Seth's metaphor and current concepts in contemporary physics. In making such comparisons, caution is essential. Seth readily concedes that he is trying to convey images of a process that is not describable in our terms — yet he must confine himself to our terms to communicate at all. He emphasizes that his descriptions are strictly metaphorical; they are not in any way an attempt at scientific explanation. In dealing with an area of physics that is not amenable at this time to experimental or theoretical confirmation, we are similarly confined to using concepts that are extrapolations from present laws, which likewise must be used with care. Even with these disclaimers, the metaphor used by Seth for the creation of space-time (matter formation) resonates with certain ideas discussed in physics. When and if the actual physics is worked out, the metaphor may be modified or completely replaced. For now we are simply searching for a useful way to intuitively picture the formation of our three-dimensional world.

Historically, the problem of perception has been viewed from two opposite poles. The philosopher John Locke (1632-1704) contended that the universe consists of solid bodies in mechanical interaction and that the objects we observe are external to us. Locke saw a causal arrangement: objects impinge on our sense organs and inform us of the nature of these external forms. George Berkeley (1685-1753), on the other hand, postulated the existence of "spirit" observers, but saw no need for any creation external to the mind. From his perspective, the experiences we have are sufficient reality in themselves. In Locke's view, then, our perceptions originate from an external universe, while in Berkeley's the perceptions themselves are the reality. While there have been many variations on these themes, Berkeley and Locke represent the two extremes.

In the paradigm being developed here, we will arrive at a view somewhere in between these opposites. We can safely say that our three observers, Bohm, Wilber, and Seth, agree that *something* is "out there." That "something" is called space-time by contemporary physicists. To Bohm it is the implicate order, to Wilber it is the deep structure, and to Seth it is Framework 2. None, however, would accept the Lockeian view that there is an objective, concrete external world causally acting on our sense organs. Rather, each feels that the world is a field of probabilities and tendencies waiting to be formed, an underlying realm of infinite energy that is passive in the sense that it does not have the ability to erupt into the physical world without an external agent. That agent is considered by our three observers to be consciousness.

According to Seth and Bohm, consciousness has a broader meaning than is normally understood. First of all, it should be noted that the agent, the passive energy, and the final physical manifestation are all the same thing. That is, they are differing forms of a unitary substance variously called the holomovement, the total ground unconscious, or All That Is — an indescribable, infinite sea of psychic energy.

Second, the agent is not in our three-dimensional world but manipulates and operates in it through the unfolding and enfolding process of the holomovement. That is, the plans for our universe, which originate in Seth's Framework 3, are used to act upon the passive energy to produce ever-changing gestalts of matter.

According to Seth, the only element that is "out there" and common to all consciousness is the passive energy. The finished product, the three-dimensional world created by each gestalt of consciousness — the electron, the cell, the human being — is an individual affair, a vehicle specifically designed for the development of that particular entity.

SETH'S CONCEPTS: A REVIEW

Consciousness Units, Electromagnetic Energy Units, and Matter Formation

According to Seth, consciousness units (CUs) move faster than light outside our three-dimensional universe and slow down to form matter. The CUs express their "mental activity" (thoughts and emotions) as electromagnetic energy (EE) units. The EE units and coordinate points form a vast subjacent energy pool from which matter emerges in a process analogous to the growing of a plant from a seed (the EE unit) with the help of the sun (the energy of the coordinate point). The EE units permeate all of space and operate just below the surface of our three-dimensional universe. Matter is their projection into our world. The EE units are not confined to human consciousness, but in varying degrees emanate from all forms, including cells and elementary particles.

Frameworks of Reality

Seth postulates a spectrum of consciousness, as do Bohm and the Perennial Philosophy, and calls each level a framework. Framework

1 is identified with our three-dimensional world. Framework 2 contains the vast assortment of probabilities available to us. Framework 3 houses, so to speak, the organizing element — in Herbert's terms, the choreographer — of our three-dimensional reality. Whether for the particle or for the human being, choices are made from Framework 3.

Probable Realities

A third concept enunciated by Seth — again, with analogs in the work of Bohm and Wilber — is that of probable events. Seth says that all consciousness (from particle to human) has an almost infinite array of probable actions to choose from within the laws of a given level of reality. Through mental means, consciousness chooses an action, thereby actualizing it in the three-dimensional world. The actions not chosen are still in contact with the consciousness, though not in the realm of our normal awareness, and may be utilized in other realities. Each selection produces more probable events for Framework 2, so that our actions in the three-dimensional universe both contribute to and draw from the pool of probabilities.

The Pulsating Nature of Matter

Seth states that all matter pulsates in and out of our universe at a rate much too rapid for our physical senses to detect. All particles, all human bodies, and for that matter, all forms of consciousness pulsate. This phasing in and out of matter does not affect our perception because the persistence of the same pattern appears as a solid body. However, in a literal sense, all particles, all objects, all beings are newly created in each moment. A particle is in our three-dimensional world for a very small fraction of its cycle of pulsation.

The Moment Point

In Framework 2, sequential time is absent; all events exist in no-time. The sequence of events as actualized in our three-dimensional universe gives rise to our notion of time. Using our terms, the past, present and future exist together in the moment point.

PHYSICS CONCEPTS

Wheeler's Space-Time: The Quantum Foam

With Seth's ideas as a background, we now turn to some ideas in modern physics to aid us in unfolding a possible metaphor for the creation of space-time. The concept of space used here is based on the work of John Wheeler. Before discussing Wheeler's theories, let us review some basic physical concepts.

After Einstein introduced the general theory of relativity, space no longer was seen as an empty field where particles cavort about; space-time itself was viewed as capable of change, a dynamic entity that can be distorted, elongated, or compressed. This dynamic potential is not limited to the expansion of the big bang, but exists wherever matter is located. A body of matter is surrounded by a gravitational field in the way that an electric charge is surrounded by an electromagnetic field and is therefore subject to dynamic forces.

It has been well known since Maxwell that if an electric charge is accelerated, electromagnetic waves are produced. If the charge is visualized as a body surrounded by a field, the movement of the body sets up a ripple that travels to the outer edges of the field at the speed of light, creating a twist in the field that travels outward from the charge. In a similar manner, an accelerated mass causes twists in space-time that travel outward and can be interpreted as gravity waves. Since Einstein showed that gravity was a deformation of space-time, this visualization seems reasonable. Gravity waves are extremely weak and difficult to detect, in contradistinction to electromagnetic waves, which are easily detected (for example, by radios).

Since quantum theory states that electromagnetic radiation is produced and absorbed in small energy packets called photons, physicists hypothesize that gravity waves have their own small distortions in space-time, called gravitons. If quantum theory is applied to gravity, Heisenberg's uncertainty principle comes into play, meaning that the graviton is subject to random fluctuations.

One of the sets of variables to which the uncertainty principle can apply is energy and time. In this case, the uncertainty is a limitation upon the accuracy of the amount of energy transferred and the time during which the transfer occurred. We cannot measure with total accuracy the energy of a given process *and* the time the process takes. The more closely we determine the energy, the less closely can we determine the time.

In effect, nature breaks the law of the conservation of energy if the period of time is short enough: the less time we have, the more energy we have. A particle can emit a second particle, which it ordinarily lacks the energy to emit, only if it can reabsorb that particle within the proper length of time. The emitted particle, which was allowed to violate the energy conservation law, is called a virtual particle. Virtual particles have not been directly observed, but their effects have been noted. As an example, an electron in orbit around a nucleus normally is expected to follow a smooth path. However, because of the brief appearance and disappearance of virtual particles, electrical fields are set up for extremely short periods of time. These fields cause slight disturbances in the electron's orbit. In 1947, W. E. Lamb and R. C. Retherford detected this disturbance in the spectrum of light from the hydrogen atom. This observation became known as the Lamb shift and is considered proof for the existence of virtual particles.

For the graviton to borrow energy in violation of the conservation law, the time must be exceedingly small, since gravity is so weak. To get enough energy to distort space-time appreciably, the time must be very, very slight indeed; we might say that gravitons can be brought into a state of existence from a state of nonexistence only if they have the good sense to go away in the proper time. In regions where this happens, space-time becomes very unstructured. What was once a vacuum is now filled with numerous different virtual particles, appearing and disappearing in amazingly short times, creating a sea of intense activity. At these very short time intervals, space-time literally tears apart with violent spurts of energy.

Wheeler describes this sea of activity as similar to a sponge with myriads of "wormholes" and "bridges" created by the violent churning below the level of particles. An elementary particle is not affected by the turmoil since it is many magnitudes larger than the wormholes. Wheeler calls the underlying level "quantum foam" and postulates that the wormholes and bridges are tears in the fabric of space-time. They could emit information into our universe and we would notice nothing unusual because the information would be lost in random collisions between atoms and molecules (see Hawking's radiation description in the section on pulsating black holes later in this chapter).

Einstein's theory of gravity ignores quantum theory. This is not of particular importance on the large scale, where quantum effects are too small to worry about, or on the particle scale, where gravitational effects are insignificant. However, well below the quantum particle level is a scale where gravity, quantum effects, and light propagation all meet, combine, and are of equal importance. In 1913, Max Planck combined the constants for gravity, the velocity of light, and the constant h (Planck's constant) and came up with natural units of length, time, and mass. The length turned out to be 1.6×10^{-33} cm, the time to be 5.3×10^{-44} seconds, and the mass to be 2.2×10^{-5} grams. It is difficult to imagine an entity the size of the Planck length: the size of earth is to the size of the nucleus of an atom as the nucleus of the atom is to the Planck length. Now, *that's small!*

A particle with this mass and length would be a black hole of quantum mechanical magnitude. Actually, the black hole can be viewed as a sphere with a radius equal to the Planck length. These quantum black holes (also called mini-black-holes) are the size of Wheeler's wormholes and bridges in the quantum foam. If space-time could be viewed at this level (which of course it cannot), it might be that quantum black holes, packed tightly together, are its basic constituents.

Planck's unit of time is independent of clocks or other such objects made by human beings, but is dependent on the way the universe is put together. This again is a consequence of Heisenberg's uncertainty principle. The shorter the time interval we measure with precision, the greater the energy concentrated in a smaller region of space. At the Planck time, the region of space becomes a black hole, thus defining the shortest possible time and smallest possible area of space. Perhaps space is a sea of virtual quantum black holes fluctuating in and out of existence at an incredible frequency defined by the Planck time.

Holes in space are actually entry and exit points for our universe. These points or holes are called singularities. A black hole singularity is where matter (or energy) leaves the physical universe, and a white hole singularity is where it enters the universe. Here is how Bob Toben and Fred Alan Wolf describe it:

> How does light get itself trapped? The answer is gravity. If the gravitational field is large enough, light gets bent by it. By bent light I mean that it no longer moves in a straight line but follows a curved one instead. The stronger the gravitational field . . . the more light bends. Finally, if the field is sufficiently strong, the light bends into a circle. And guess where gravity gets that strong. Did you guess it? Yes, indeed, in a black hole.

> Following on Wheeler's idea of quantum foam . . . consisting of spontaneously occurring miniblack holes and white holes, we may envision this foam as entrapping light in its bubbles. These tiny gaps in space-time generate very strong gravitational fields over very, very small distances. At the center of these bubbles are points called singularities, singular points in space-time where forces are beyond comprehension. It is because of these tiny "beyond space-time" points that gravity can entrap light.[1]

In corroboration, Jack Sarfatti says:

> The bubbles that continually appear and disappear in the geometry of space are miniblackholes and miniwhiteholes. Each can be of either positive or negative mass. The miniblackholes are virtual particles; the miniwhiteholes are virtual anti-particles. Virtual means that the particles have only a transient existence, being continually created and destroyed.[2]

Both Wolf and Sarfatti say that when we consider these elemental regions of space-time as defined by Planck units, the quantum energy can become so high that it can bind itself together temporarily and become a quantum black hole.

Since our physical laws are completely integrated with our conception of space-time, and since space-time literally self-destructs at a singularity, our known laws are of little value in this region. Prediction becomes impossible.

In our discussion of the measurement problem in physics (Chapter 3, Matter Formation: A Physics Metaphor, p. 131), we saw that to describe an electron's motion from one point to another, the physicist employs a wave function that gives the probabilities of the different trajectories that the electron might take. The actual trajectory upon measurement collapses the wave function, and one solution is the result. Before the measurement takes place, all possible trajectories coexist in a potential fashion. Hugh Everett's interpretation of this phenomenon assumes that all possible trajectories actually exist in a gigantic multidimensional superworld, and the one we select is what is expressible in our three-dimensional world. In an analogous way, the graviton is guided by a wave function, and since the graviton is a distortion of space, each probability for a graviton represents a different shape to space. All these probabilities together constitute what Wheeler calls superspace. Our three-dimensional world, then, is only one projection out of this superspace. And, as Sarfatti says, each singularity connects different points in superspace.

Elementary Particles

When we consider how physicists view an elementary particle, we have to say that they view it with great difficulty. Paul Davies gives an excellent description of the problem in his book *Superforce*. The following discussion is based on his approach.

Let us assume that the elementary particle is an electron in space. The space itself is seething with virtual activity so that the electromagnetic field that surrounds the particle becomes a cloud of energy. If it were possible to send a probe into this cloud, we would find that the amount of energy rapidly rises as the electron is approached. In fact, when the energy cloud is computed, it turns out to be infinite unless the electron itself can be considered to be larger than a point. To avoid the problems raised by infinite energy, the electron might be seen as a small rigid entity like a tiny billiard ball. But this also presents a problem since it contradicts relativity theory. Information cannot be transferred any faster than the speed of light. If a force were applied to one side of the electron ball, the entire ball could not be set in motion at the same time because all parts of the ball would not receive the information of the applied force instantaneously. Therefore, it could not remain rigid. Yet if it were not rigid, the internal repellent electric forces would blow the electron apart. This has not been observed, and it would seem that the particle is neither a point nor a rigid body.

As Davies points out, the mass of a measured electron is made up of two parts, the so-called bare electron, which is at the center, and the electrical energy surrounding it. The observed mass is finite, but a part of the observed electron, the surrounding field, computes to be infinite. Through a mathematical device called renormalization, the infinity portion can be removed, and then the theory seems to agree with experimental evidence. Since infinities do not show up in measurements, an infinite term is manageable mathematically only as long as it does not refer to a measurable quantity. In nongravitational physics, only energy *differences* are measured. Hence, one is free to rescale the infinite term by an infinite amount.

From the point of view of quantum electrodynamics, the electromagnetic force is mediated by a photon. We have seen that according to Heisenberg's uncertainty principle, an electron can emit a photon and reabsorb it in the proper time without violating conservation laws. This virtual photon contributes to the mass of the electron. If the electron is a point, the time to issue a photon and return it can go to zero, with the resulting energy going to infinity. Since the point electron can emit any number of photons, the solution is an unending series of infinite terms.

This problem of the difference between the measured and the calculated values for an electron attribute took twenty years to resolve, finally making quantum electrodynamics usable. The issue revolves around the electron being a point. If the electron could in some way be enlarged, the infinities *might* disappear. Present-day physics deals with dimensions down to 10^{-15} cm but not below. At smaller dimensions, gravitational effects come into play, and space-time becomes distorted. Whether these distortions could eliminate the infinities is unknown. Recall that Bohm was looking for electron "structure" between 10^{-15} cm and 10^{-33} cm. The whole procedure of renormalization revolves around the fact that physicists know very little about what occurs at these extremely small dimensions. However, since renormalization works so well, most physicists accept it and use it, though a few, like Richard Feynman, question it as a legitimate procedure.

No matter how clever the word [renormalization], it is what I would call a dippy process! Having to resort to such hocus-pocus has prevented us from proving that the theory of quantum electrodynamics is mathematically self-consistent. It's surprising that the theory still hasn't been proved self-consistent one way or the other by now; I suspect that renormalization is not mathematically legitimate.[3]

And David Bohm says:

> With regard to such problems, we first note that the present relativistic quantum field theory meets severe difficulties which raise serious doubts as to its internal self-consistency. There are the difficulties arising in connection with the divergences (infinite results) obtained in calculations of the effects of interactions of various kinds of particles and fields. It is true that for the special case of electromagnetic interactions such divergences can be avoided to a certain extent by means of the so-called "renormalization" techniques. It is by no means clear, however, that these techniques can be placed on a secure logical mathematical basis. . . . While it has not been proved conclusively, as yet, that the infinities described above are essential characteristics of the theory, there is already a considerable amount of evidence in favour of such a conclusion.[4]

Bohm feels that since renormalization may not be mathematically legitimate, a fundamental change in present theory may be required when treating interactions at these incredibly small distances.

Black/White Holes

Black holes are unsettling for physicists because within them, all physical laws go up in smoke. The black hole curves space and warps time. In simple terms, it is a hole in space where whatever falls in never comes out.

The concept of the black hole was introduced by astronomer Karl Schwarzschild in 1916. Schwarzchild pushed Einstein's concept of curved space-time to the extreme position: if enough matter is gathered into a small enough volume, the space around it takes on an infinite curvature, which is the black hole. Schwarzschild worked out the mathematics involving the relationships between size, mass, and diameter of celestial bodies needed to produce a black hole. This work became known as Schwarzschild's solution to Einstein's equations. Almost half a century later, in 1963, physicist Roy P. Kerr revolutionized black hole theory by adding rotation to the previously motionless hole. Since the black hole approaches infinite density, the rotation is extremely rapid. Most physicists agree that during the period in which the hole absorbs matter, it must be rotating.

Since Einstein's equations are time-symmetrical, they provide for white hole solutions as well as black hole solutions. The white hole is the exact opposite of the black hole. A black hole implodes, and a white hole explodes. Everything is trapped in a black hole; everything goes out of a white hole. Anything that occurs in a black hole also occurs in a white hole, but in reverse.

Several physicists have observed that a black hole can be seen as a time machine. That is, to journey (hypothetically) through the black hole is to go backward in time. Conversely, a journey through a white hole is to go forward in time.

In 1935, Einstein and Rosen used the Schwarzschild solution as a model for an elementary particle, which they attempted to describe as a twist in space-time. While their model is not generally used, they did make a remarkable discovery. By extending the Schwarzschild solution, they found that the hole in space opens up and connects to a second or parallel universe. This connection is called an Einstein-Rosen bridge. Since experimental evidence of a black hole is not available, the interpretation of this second universe is open to question. Einstein felt that it was a pathway to a distant part of our own universe. At each end he envisioned a particle with opposite charges. Others feel that Einstein's interpretation is not valid, at least for small black holes, and that the second universe is "outside" our own. Since the full solution of Einstein's equations requires a white hole, the second universe would also require a black hole – white hole pair.

SETH'S SPACE AND
INVISIBLE (VIRTUAL) PARTICLES

According to contemporary physics, space is a vast sea of virtual particles that appear and disappear in incredibly small segments of time. This creates a turbulent energy pool — Wheeler's quantum foam — that literally tears apart the fabric of space-time, giving it (according to Wheeler) a spongelike texture of minute wormholes and bridges. All this activity takes place well below the magnitudes of elementary particles. Seth's image of space is similar:

Space as you think of it is . . . filled with invisible particles. They are the unstated portion of physical reality, the unmanifest medium in which your world exists. In that regard, however, atoms and molecules *are* stated [explicated], though you cannot see them with your eye. The smaller particles that make them up become "smaller and smaller," finally disappearing from the examination of any kind of physical instrument . . . [5]

Seth uses the idea of black holes and white holes to describe the pulsations of the CUs, expressed in EE units, in and out of the universe. He sees matter as projections into our universe when a persistent pattern of EE units results in form. Pulsations become matter through forms of elementary particles, which in turn follow forms that create atoms, molecules, and so forth. Space on the other hand, is a seething cauldron of EE units without form, and which only momentarily breaks through and does not create matter. In Seth's words:

This inward and outward thrust allows for several important conditions that are necessary for the establishment of "relatively" separate, stable universe systems. Such a system may seem like a closed one from any viewpoint within itself. Yet this inward and outward thrusting condition effectively sets up the boundaries and uniqueness of each universal system, while allowing for a constant give-and-take of energy among them.

No energy is ever lost. It *may* seem to disappear from one system, but if so, it will emerge in another. The inward and outward thrust that is not perceived is largely responsible for what you think of as ordinary consecutive time. It is of the utmost and supreme importance, of course, that these CU's are literally indestructible. They can take any form, organize themselves in any kind of time-behavior and seem to form a reality that is completely dependent upon its apparent form and structure. Yet, [when energy disappears] through one of the physicists' black holes, for example, though structure and form would seem to be annihilated and time drastically altered, there would be an emergence at the other end, where the whole "package of a universe," having been closed in the black hole, would be reopened.

There is the constant surge into your universe of new energy through infinite minute sources. The sources are the CU's themselves. In their own way, and using an *analogy*, now, in certain respects at least the CU's operate as minute but extremely potent black holes *and* white holes, as they are presently understood by your physicists.

The CU's, following that analogy, serve as source points or "holes" through which energy falls into your system, or is attracted to it — and in so doing, forms it. The experience of forward time and the appearance of physical matter in space and time, and all the phenomenal world, results. As CU's leave your system, time is broken down. Its effects are no longer experienced as consecutive, and matter becomes more and more plastic until its mental elements become apparent. New CU's enter and leave your system constantly . . . .

The EE units . . . represent the stage of emergence, the threshold point that practically activates the CU's, in your terms.[6]

The EE units, when they operate as white holes, bring energy into our universe, and when they operate as black holes, take energy away. There apparently is a conservation of energy law for All That Is, since no energy is ever lost in this process. The EE units are indestructible.

At this point, Sarfatti and Seth differ slightly. Sarfatti identifies the virtual particle as either a black hole or a white hole. Seth says the black hole and the white hole are within each other. When energy is being absorbed, it is a black hole; when it is being released, it is a white hole. Again, in Seth's words:

Using this analogy of the white hole and the black hole: to make this clearer, the white hole is within the black hole.

Electromagnetic properties are drawn into the black hole, and accelerated beyond imagination. The acceleration and the activities within the black hole draw unbelievable proportions . . . of additional energy from other systems.

This greater acceleration changes the very nature of the units involved. In the meantime, the characteristics of the black hole itself are changed by this activity. A black hole is a white hole turned inside out, in other words. The electromagnetic "matter" may reemerge through the same "hole" or "point" which is now a white hole.

The reemergence, however, again alters its characteristics. It becomes "hungry" once more, and again, a black hole. The same sort of activity goes on in all systems.[7]

In effect, the hole goes through a cycle of absorbing energy as a black hole and disgorging energy as a white hole.

The idea of these holes as being exit and entry points for a number of universes coincides with Wheeler's concept of superspace. If one extends the thought of probable universes to the graviton, the three-dimensional world is seen as only one aspect of a superworld. Since the graviton is guided by a wave function, and since the graviton represents a distortion, or wrinkle, in space, each probability for a graviton represents a different shape to space. All these together make up Wheeler's superspace, a gigantic multidimensional superworld. Sarfatti says:

> The idea that reality is composed of an indefinite number of coexisting universes is expressed in modern physics by Wheeler's concept of superspace. The singularities of space-time are exit and entry points connecting these different universes.[8]

In Seth's terms, these connecting points are the EE units, acting as conduits to Framework 2. Energy is brought into our universe and taken out by a process analogous to a pulsating black/white hole.

ELECTROMAGNETIC SPECTRUM: A COMMON VELOCITY

In our three-dimensional world, electromagnetic energy is propagated by a vibrating electric and magnetic field created by a moving charge (such as an electric current). Since the field is vibrating, it has a frequency. In fact, there exists a whole spectrum of electromagnetic fields that vary with frequency. What we call visible light is only a small portion of the entire spectrum. For the sake of our discussion here, the entire spectrum will be referred to as light. There are no sharp boundaries between the various frequencies of radiation, and all the frequencies of the spectrum have one thing in common: all travel at a maximum velocity referred to as the velocity of light.

Einstein's famous mass-energy equation, $E=mc^2$, the result of special relativity theory, states that the total energy of a mass (including rest mass, kinetic energy, and potential energy) is equivalent to the mass multiplied by the speed of light squared. The theory also states that the speed of light in a vacuum is constant for all observers regardless of the motion of the source or of the observers. This, of course, is contrary to common sense but is totally accepted since it is borne out empirically, the most famous instance being the Michaelson-Morley experiment. In our universe, then, the constancy of light's velocity indicates that as mass increases or decreases, energy follows suit in a direct fashion.

Physicists presently describe the world through mathematical laws that utilize what are called dimensionless numbers, that is, numbers not expressed in terms of arbitrary standard units. These numerical values, therefore, may have significance, and if any one of these numbers is changed, the universe is changed. This suggests that there could be a group of universes, each with different values of the dimensionless numbers. Some physicists have noted cases of relationships between these numbers that do not appear to be arbitrary but are unexplainable. These dimensionless numbers are established from experimentally determined constants, which in themselves are not dimensionless but which determine the value of the dimensionless numbers. Such physical constants are sprinkled throughout the laws of physics and determine the unique features of our universe.

The fact that these fundamental constants must be discovered through empirical means rather than theoretically indicates to some physicists that the theory is incomplete in some way. As an example, the mass of the electron in quantum theory is determined in the laboratory and then inserted in the appropriate equation. It does not emerge in a natural way. The same is true for electron charge in field theory and the velocity of light in relativity theory. In the case of the velocity of light, there is a formula for its determination, but this formula is dependent on two experimentally determined constants. If the velocity of light changes, the universe changes. If there are a number of possible universes, as this might suggest, the velocity of light defines our universe uniquely.

That the universe is constrained by the velocity of light is true whether the source and the observer are moving in the same direction or in opposite directions. This fact is the cornerstone of Einstein's special theory of relativity; from this flows all the consequences of the theory. Our concepts of space and time must be adjusted to accommodate this phenomenon. Thus, physicists join space and time together into a four-dimensional space-time interval that remains constant for all observers. That is, space and time separately are varied to keep this interval constant.

Seth talks about EE units slowing down to our velocity of light in order to create matter.

> Your light . . . represents only a portion of an even larger spectrum than that of which you know; and when your scientists study its properties, they can only investigate light as it intrudes into the three-dimensional system.[9]

As Seth sees it, our electromagnetic spectrum is only a portion of an infinite spectrum in which each portion is defined by the velocity of the EE units. Our light has a velocity of 300,000 km/sec, and that velocity defines our universe and our reality. Can we speculate that there are an infinite number of realities, each having a particular velocity for its electromagnetic spectrum? Seth says yes! He calls these realities code systems and describes them as follows:

> Consciousness operates with what you might call code systems. These are beyond count. . . . These code systems involve molecular con- structions and light values, and in certain ways the light values are as precisely and effectively used as your alphabet is. For example, certain kinds of life obviously respond to spectrums with which you are not familiar — but beyond that there are electromagnetic ranges, or rather *extensions* of electromagnetic ranges, completely unknown to you, to which other life forms respond.[10]

We focus within our code system and block out other code sys- tems when operating with our normal senses. The notion of a large number of code systems finds common ground with Bohm's views. In an interview with Weber, the following exchange took place:

> *Weber:* Our kind of space and time is one among perhaps infinitely many orderings possible in the universe yet we think that it's the only way and in fact the necessary condition for understanding. Kant almost said that and could not conceive of an alternative arrange- ment. The super-implicate order proposes an alternative to current narrow western epistemology.

> *Bohm:* Yes, it says that the information content out of which the implicate order unfolds is not determined in an order of space and time as we know it, but it contains that order within it.[11]

Weber and Bohm are saying that the implicate order contains within it an infinite number of possible orders, or code systems, to use Seth's term.

Einstein's special theory of relativity offers us another way of looking at other possible code systems. The theory states that time can be seen as equivalent to a spatial axis if the time interval is multiplied by the velocity of light. The time axis turns out to be negative and therefore can be understood as an imaginary dimension. Because the velocity of light is large, one second of time is equivalent to 300,000 km of space. Using this metaphor, the passage of time means that we are actually moving down the time axis at the velocity of light. Why, then, do we feel that we are at rest while reading this book? Because everything in the universe is moving along the time coordinate at the same velocity. Relatively speaking, we are at rest.

Since the velocity of light determines the amount of energy within a given mass, it should affect the subtlety of matter within our code system. The higher the velocity, the more subtle the mass. Seth says that since we focus on a particular code system defined by our velocity of light, other systems can exist in the same space, and we would not be aware of their particular masses since our normal senses would not detect them. Realities are not separated by space, then, but by the velocity of light.

Descriptions of the mystical experience and other altered states of consciousness often characterize these other systems as involving another frequency domain, usually a higher frequency. This experience might be related to Seth's higher-velocity systems (see pp. 133, 213) and possibly to the following ideas in contemporary physics. The uncertainty principle states that when a particle of a mass m behaves like a wave and travels at close to the velocity of light, it will have a wavelength equal to h/mc. This relationship indicates that the wavelength of the mass is inversely proportional to the velocity of light, and the frequency is directly proportional to the velocity of light. In other words, the higher the velocity of light, the higher the frequency domain.

A description that conforms to this concept of various domains operating at different frequencies is found in the near-death experience related by architect Stefan von Jankovich:

One of the greatest discoveries I made during death . . . was the oscillation principle. . . . Since that time "God" represents, for me, a source of primal energy, inexhaustible and timeless, continually radiating energy, absorbing energy and constantly pulsating. . . . Different worlds are formed from different oscillations; the frequencies determine the differences . . . . Therefore it is possible for different worlds to exist simultaneously in the same place, since the oscillations that do not correspond with each other also do not influence themselves. . . . Thus birth and death can be understood as events in which, from one oscillation frequency and therefore from one world, we come into another.[12]

In summary, Seth says that our three-dimensional reality is defined by our velocity of light. This is the velocity that EE units slow down to when entering our universe. According to contemporary physics, our light velocity determines the frequency domain we occupy and the sublety of the matter we create.

STRINGS, AN ALTERNATIVE METAPHOR

Superstring theory, put forth in recent years, offers an alternative to the black/white hole metaphor and combines two elements that we have discussed. First, it assumes that the universe has more than three dimensions and that these extra dimensions are hidden. Second, it assumes that particles are not points with all of their attendant infinities, but rather that the basic constituents of the universe are tiny vibrating strings. Theoretically these strings can be open or closed, but the hypotheses that show the most promise propose closed strings. The size of these strings is in the region of the Planck length, 1.6×10^{-33} cm. Thus they are similar in dimension to the postulated black/white holes.

According to physicist Ed Witten, there is basically one kind of string, which, like a violin string, can vibrate at any number of frequencies. These frequencies, or harmonics, have unique energy levels, which correspond to the different masses of various elementary particles. Strings exist in a higher-dimensional world, beyond our three dimensions of space and one of time. The additional dimensions are not observed because they are "curled up" in diameters of 1.6×10^{-33} cm. In effect, this means that the universe has hidden dimensions.

If we could view a string with a powerful enough microscope, would it appear as a wriggling particle? Physicist Michael Green answers as follows:

> The theory is probably much deeper than that, because just as one is looking at those scales on which one would see the wriggles, so to speak, one is then looking at the scales in which the whole structure of space and time has to be modified. Therefore it may not be correct even to think of this thing as moving through what we normally think of as continuous space and time.
>
> The notion of a string is inseparable from the space and time in which it's moving, and therefore if one has radically modified one's notion of the particle responsible for gravity, so that now it's stringlike, one is also forced to abandon at some level the conventional notions of the structure of space and time. When I say at some level, the level I'm talking of is at these incredibly short scales associated with the Planck distance.[13]

Green says that besides vibrating, these strings have structure. Different charges can reside on the string, and "the nature of these charges . . . distinguishes different particles."[14]

The concepts of mini-black/white-holes and perhaps vibrating strings suggests that the basic constituents of matter reside in the 10^{-33} cm region. (It should be noted that at this time, string theory does not describe black holes.) Seth places his CUs in the same region. Whether the basic units will be viewed as black holes remains to be seen. At present, black holes are not susceptible to experiment because all *known* laws of physics break down at singularities. But this may not always be the case. When Einstein developed his theory of relativity, he discarded Newton's concept of an absolute space and time, but he did not discard the idea of continuous space and dimensionless points. According to relativity theory, space is still considered continuous and infinitely divisible.

One possibility suggested by quantum theory is that space-time is a plenum rather than a void. A second possibility suggested by quantum theory is this paradox: Schrödinger's wave equation is based on calculus (differential equations), which postulates a continuous space wherein certain properties can be infinitesimally close, but Heisenberg's uncertainty principle states that space cannot be divided into units that are infinitely small. This paradox might be resolved by the notion that space is actually created from an underlying level. If this subjacent level is nonlocal, a nonlocal description of black holes might be possible.

The singularity is now considered a single point in space, and space-time breaks down at the point because of the infinities that arise. However, if space-time is a second-level phenomenon, the singularity might be included with space-time once its properties are derived. When a point in quantum theory is thought of as spread out, and, in a sense, nonlocal, then infinities may be avoided. Space-time may turn out not to be the womb where elementary particles are created, but space-time itself may be created out of quantum processes on a subquantum level.

In any case, in string theory the electron is seen not as a point but as having structure. Green sums this up:

> One way of describing quantum mechanics is in terms of the so-called uncertainty principle and by using the uncertainty principle, it's easy to argue that the shorter the distance scale that you're trying to describe, the more uncertainty there is in the energy of what you're trying to describe. Now in a theory of gravity this means that when you try to describe things at incredibly short distances (and by short I mean *unbelievably* short compared to the size, say, of even the proton), the fluctuation in energy of what you are looking at might be big enough to make a little black hole. So if we are contemplating making an observation on small enough distance scales (this scale is called the Planck distance, it is 10^{-33} centimetres), we are forced to think of even empty space as consisting of an infinite sea of *fluctuating* * black holes, coming and going in very short times. This of course radically alters our notions of what we mean by space, and it's a disaster because we don't really understand what's happening any more. The whole notion of space itself as being made of points probably no longer makes sense.[15] [*Emphasis added: the fluctuating, or pulsating, black hole is discussed in the following section.]

The proponents of string theory hope that the basic constituents of the universe will prove to be these tiny vibrating strings, thus fulfilling the quest for the ultimate level of reality. Of course, with the view taken here, the string could be a gateway to other realities. David Peat observes:

It is possible that as we probe the elementary particles [and] super-strings . . . they may not lead to a single most fundamental level, but rather they may open into a world of ever greater richness and subtlety. Instead of the end of physics being in sight, science may enter a new realm of complexity and new forms of order that con-tain, for example, feedback loops from much higher levels and even interconnections with the large-scale structure of the universe it-self. Rather than being *the theory of everything*, superstrings may be the door to another universe.[16]

We have seen in these last three sections that as gravity becomes strong enough on the quantum level, space and time literally come apart. Wheeler sees this as analogous to a solid that has reached its elastic limit. The substructure exposed by stretching space-time to its limit is labeled by Wheeler pregeometry. The black hole is the gate-way to the world of pregeometry. Does that limit represent a doorway to other realities or just an entrance to chaos? Peat says that "all of nature is contained within a single element of space and time . . . ."[17] If that element is defined by the Planck length and time, then all of space is analogous to a hologram, and each single element is a door-way to the implicate order, or, in Seth's terms, to Framework 2. Seth sees the physicist's hand on the doorknob, but the door as yet un-opened. Whatever is beyond the door, the door itself seems to be the entryway for space-time, and its size is in the region of the Planck length.

THE PULSATING BLACK HOLE,
A COMMON METAPHOR

Seth describes the elementary particle as having a mini-black/white-hole at its core, surrounded by faster- or slower-than-light par-ticles, depending on whether it is in the black hole or white hole phase. Paul Davies describes an electron as being composed of two parts, the "bare mass" at the core (minus the electromagnetic field) and the sur-rounding shroud of fluctuating quantum energy made up of all man-ner of virtual particles. He says:

The electromagnetic field which clothes the particle must be viewed as a retinue of virtual photons fussing around the electron, clinging to it, and forming a tenacious shroud of energy. . . . Those that remain close to the electron, near the center of the shroud, carry considerable energy; in fact, when the total energy of the photon shroud is computed, it again turns out to be infinite.[18]

Wheeler comments as follows:

> If there is any correspondence at all between this virtual foam-like structure and the physical vacuum as it has come to be known through quantum electrodynamics, then there seems to be no escape from identifying these wormholes with "undressed electrons." Completely different from these "undressed electrons," according to all available evidence, are the electrons and other particles of experimental physics. For these particles the geometrodynamical picture suggests the model of collective disturbances in a virtual foam-like vacuum, analogous to different kinds of phonons or excitons in a solid.[19]

Instead of using the term "bare electron," Wheeler uses the term "undressed electron." The equivalence is apparent.

In the black hole phase described by Seth, the electron core is surrounded by faster-than-light particles, the counterpart of the photon shroud that clothes the electron. During this phase, the core becomes heavier and heavier and sends its particles out of our universe. In a sense, the core collapses on itself, or falls through a hole in space, or seems to swallow itself. The faster-than-light particles follow the collapse and act like a lid to the black hole, which gradually diminishes. As far as our universe is concerned, the hole is closed at this point. As the core goes out of our universe, the particles of the core accelerate while the original faster-than-light particles decelerate. On its return phase, the core is now faster than light, and the surrounding particles are slower. The former black hole then emerges as a white hole in our universe. Energy is released through its core, beginning the black hole phase once again.

The concept of a black hole cycling in and out of existence has a counterpart in physics in the big bang theory and the cycling universe. That is, at one time the universe was contained in the initial black hole and then exploded outward to form space and time. If the universe is cyclical, it will return to the primeval black hole and start all over again. Of course, the time cycle involved is far different from the Planck time of 5.3×10^{-44} seconds.

In 1974, Stephen Hawking found that according to quantum theory, the black hole emits particles at a steady rate. This discovery was astounding, since conventional wisdom held that matter could not escape from a black hole. In addition, a black hole that emits particles steadily decreases in mass and size. After all, energy must be balanced. The energy to create the particle must come from the black hole, or the gravitational field of the hole. That means that the mass of the hole must decrease and evaporation must occur. For large black holes, say with the mass of our sun, the time needed to disappear would be incredibly long: 10^{66} years. Hawking also postulates what he calls "primordial" black holes, which he believes may have existed shortly after the big bang. These would have a radius of 10^{-13} cm and would evaporate in 10 billion years. While this seems a long time, it is relatively short compared with 10^{66} years for the sun. However, when we talk about black holes with a radius of 1.6×10^{-33} cm, the evaporation time is in the range of the Planck time. In short, in a black hole, matter collapses and is lost, but new matter is created in its place. Of course, the confirmation of our treatment of quantum black holes will not be possible until we at least have a consistent theory of quantum gravity.

The creation of new matter makes black holes on a subquantum scale look like white holes. This new matter contains new information and can therefore be considered an information source. Hawking has suggested that black holes and the time-reversed white hole are physically identical. That is, in some fashion the absorption of material by the black hole is the time-reverse of the quantum radiation he discovered. It should be pointed out that Hawking's evaporation process is considered random. Because of the uncertainty principle, the prediction of the appearance of a particle ejected from a black hole is not possible; therefore, the information coming out of a black/white hole is random. All the information literally pouring into our universe must be unpredictable. As Hawking points out, "God not only plays dice but also sometimes throws them where they cannot be seen." But remember our thesis that randomicity merely indicates a restricted context.

Seth describes the core of these holes as a vortex. The implication is that the core rotates rapidly around its center and acts like a suction in its black hole phase to draw the faster-than-light particles into the hole, which then closes. In its white hole phase, the core is also a vortex, except that its center exhibits faster-than-light particles surrounded by slower-than-light particles. If the core is made up of EE units, it will exhibit a magnetic field perpendicular to the mouth of the hole.

Seth's description of the core as being surrounded by faster-than-light particles might be compared to the ergosphere found in rotating black holes. If one could visualize space as a whirlpool that is both rotating and flowing into the black hole simultaneously, two surfaces are created, called event horizons, which enclose the black hole. The ergosphere is the region between the outer event horizon (also called the static surface) and the inner event horizon. It consists of swirling space that is spiraling inward to the inner event horizon. All matter found within this vortex is carried around in a whirlpool of space. Furthermore, it is transported faster than light relative to a distant observer. The inward component of the velocity is less than the velocity of light. Special relativity theory and its velocity of light restriction apply only to matter in space, not to space itself. As an example, there are parts of distant space that expand faster than light. In his book *Cosmology*, Edward Harrison notes:

> Special relativity is still valid in small local regions; for instance, the local speed of light that we use, and nothing can move through space faster than light. But on the large scale, space is so deformed that we can regard it as flowing inward with a speed that has no limit, and small local regions — in which special relativity is valid — are carried along with it. If we were in a spaceship that fell into a black hole, we could never escape; no matter how strong the thrust of the rocket engines, we could not exceed the speed of light in local space and would therefore be carried to our doom.[20]

Harrison is saying that the restrictions on the velocity of light are a consequence of the special theory of relativity, which assumes a flat space-time. However, when space is curved, light can violate the limit.

Another property of the ergosphere is relevant here. As the rotation rate of the black hole increases, the outer event horizon shrinks. At the same time, the inner event horizon grows. At a maximum spin, the ergosphere goes to zero, and we are left with a naked singularity. This might be compared to the "lid" stage described by Seth (see p. 219). Also, the singularity left by a rotating black hole is not a point, it is a ring.

SPIN AND TIME: PROJECTION AND INJECTION

Lawrence Leshan and Henry Margenau mention the implications of the strange things that can happen within black holes:

> The metric (geometry) of space-time is altered; time and space interchange their roles. To some theorists this suggests that conscious beings inside a black hole can go back and forth in time, reliving and preliving their life experiences.[21]

In this connection, Seth briefly brings up the subject of electron spin but has great difficulty explaining it.

> In as simple language as possible, and to some extent in your terms, the electron's spin determines time "sequences" from your viewpoint. . . . Time, in your terms then, is spinning *newly* backward as surely as it is spinning newly into the future. And it is spinning outward *and* inward into all probabilities simultaneously.[22]

Spin in the quantum world has several peculiar properties. First, quantum theory tells us that an electron can spin in only two directions, "up" or "down." (The meaning of this concept is vague. Suppose there is only one electron in the universe: what difference is there between up or down if there are no reference points?)

Second, some elementary particles exhibit what is known as the double rotation property. The quantum physicist divides the world into two classes of particles, bosons and fermions. Bosons exhibit wavelike rather than particulate properties. They can occupy the same space and time, they can interfere like waves and produce a single effect, and they seem to bunch together though no attractive forces are involved. The fermions, on the other hand, cannot occupy the same quantum state and are thus able to produce atoms. This property is the result of Pauli's exclusion principle[23] and cannot be explained classically. The fermions are the constituents of matter, while bosons are the constituents of force. All particles are said to have spin, the fermions differing from the bosons in that they have half-integral units of spin and exhibit the double rotation property.

Since the particle is seen as having no dimensions, double rotation cannot be understood by envisioning a spinning ball. Electron spin simply reflects the mathematical transformation of the electron's wave function (which seems to be analogous to the mathematical description of a spinning ball).[24] Double rotation, the result of combining relativity and quantum theory, means that the electron has to turn around twice to get back where it started. Picturing this accurately is probably not possible. However, for purposes of analogy only, imagine the electron spinning about its axis clockwise. Then tip the electron end over end so that it is spinning counterclockwise. Thus it has turned 180 degrees. Now turn it another 180 degrees. Lo and behold, it does not return to its original position. For a rotated electron to have its original state restored to that of a nonrotated electron, the rotation must go through two complete revolutions, or 720 degrees. Davies describes the consequence of this:

> Evidently, in the primitive case, a rotation of 720 degrees is necessary to produce a complete revolution, i.e. to restore the world to its original configuration. An elementary particle, such as an electron, perceives the total sweep of 720 degrees. In human beings, and other large objects, this facility is lost, and we cannot distinguish one 360 degree rotation from the next. In some sense, then, we perceive only half the world that is available to the electron. . . . The curious "double image" view of the world possessed by electrons and other quantum particles, is considered to be a fundamental property of nature.[25]

What is the half of the world that we cannot perceive? Seth says that the electron "is spinning outward and inward into all probabilities simultaneously." He defines spin as the cycling of the particles in and out of our universe. Seth further says that an elementary particle has a core that is a black hole, which cycles into a white hole and then again into a black hole. If the particle is a fermion, 360 degrees could be the period for the black hole to turn into a white hole, while a second 360 degrees turns it back into a black hole. So the complete cycle is 720 degrees for the particle to get back to its original position. If a positron-electron pair are to be exchanged — each to turn into its opposite — then each one needs to be rotated 360 degrees relative to the other, a total of 720 degrees. Perhaps we can identify Seth's black hole with a positron and the white hole with an electron.

The positron is the antiparticle of the electron. The concept of antiparticles arose when physicist Paul Dirac, combining quantum theory and the principle of relativity, deduced an equation for an electron wave that has two solutions: the first describes the electron, and the second describes an electron with a positive charge, or antiparticle, the positron. This equation led to the more general principle of antimatter.[26] In essence, all the distinguishing quantities of a particle except mass are reversed when matter becomes antimatter. More important, matter and antimatter go hand in hand. Physicist Heinz Pagels says:

> If this mathematical description of the creation and annihilation of quantum particles is carried out in the context of relativistic quantum-field theory, one finds that one cannot have the possibility of creating a quantum particle without also having the possibility of creating a new kind of particle — its antiparticle. The existence of antimatter is simply forced on one by . . . the creation and annihilation process in accord with both quantum and relativity theory.[27]

Regarding Dirac's discovery, Heisenberg commented:

> I believe that the discovery of particles and antiparticles by Dirac has changed our whole outlook on atomic physics. . . . Up to that time I think every physicist had thought of the elementary particles along the lines of the philosophy of Democritus, namely by considering these elementary particles as unchangeable units which are just given in nature and are always the same thing, they never change, they never can be transmuted into anything else. They are not dynamical systems, they just exist in themselves. After Dirac's discovery everything looked different, because one could ask, why should a proton not sometimes be a proton plus a pair of electron and positron and so on? . . . Thereby the problem of dividing matter had come into a different light.[28]

Today it is accepted that matter is not stable forever but can be created and destroyed. In fact, matter can be viewed as condensed energy. In the beginning (at the time of the big bang), only energy was in the universe. Matter came later as a concentration of the energy of the big bang. However, when matter is created in the laboratory, for each particle created there is a corresponding particle of antimatter. For the electron, this is the positron with the same mass as an electron but with an opposite charge and with the spin reversed. When an elementary particle collides with its antiparticle, all of the condensed energy is released, and both particles are annihilated. Thus, there cannot be a significant amount of antimatter in the universe since it would completely destroy the matter of the universe.

The question is, then, how was matter created without creating antimatter? The only answer suggested thus far (for physicists) is offered by the speculative grand unified theories (GUT). According to GUT, at the beginning a slight asymmetry developed between matter and antimatter, with the difference being in matter's favor. As the universe cooled, all the antimatter and a corresponding amount of matter was destroyed. The small excess of matter went on to form the stars and galaxies of our universe. The only antimatter seen by physicists is that which is created by physicists. No destruction of matter and antimatter has been observed outside the laboratory (with the exception of cosmic ray showers).

MATTER, ANTIMATTER, AND INTERVALS: A CYCLING BLACK HOLE

About intervals and nonintervals Seth says:

For every moment of time that you seem to exist in this universe, you do not exist in it. The atoms and molecules have a pulsating nature that you do not usually perceive, so what seems to you to be a continuous atom or molecule is, instead, a series of pulsations that you cannot keep track of.

Physical matter is not permanent. You only perceive it as continuous; your perceptive mechanisms are not equipped to detect the pulsations.[29]

The noninterval is when matter does not exist, and the interval is when it does. This pulsation can be considered the cycling of the black/ white hole. If the cycling corresponds to the physicist's rotation of a spinning particle, then the black hole for an electron is when it is in its positron phase and is going backward in time. Seth makes an interesting statement after describing the pulsation process: "The same applies to negative or antimatter however, which you do not perceive in any case."[30] In effect, we do not experience an antimatter universe because energy is not sent into our universe during this stage.

The cycling is clearly demonstrated through the use of Feynman diagrams, in which the lifeline of a particle is plotted on a two-dimensional coordinate system, the vertical axis being time and the horizontal being space. One of the amazing results of this approach is described by physicist John Gribbin:

> The dramatic discovery that Feynman made in 1949 is that the space-time description of a positron moving forward in time is *exactly* equivalent to the same mathematical description of an electron moving backward through time along the same track in the Feynman diagram.[31]

If all particle pulsations are in phase, the whole universe is either matter or antimatter at the same time. It is matter when the white hole is in our universe. When we are in the matter phase, whatever universe (or universes) is on the other side of the black/white hole is in the antimatter phase. Seth says that these nonintervals are "moments in other dimensions of reality."[32] That is, when our universe is "turned on," our sister universe is "turned off." Sarfatti comments:

> The bubbles that continually appear and disappear in the geometry of space are miniblackholes and miniwhiteholes. Each can be of either positive or negative mass. The miniblackholes are virtual particles; the miniwhiteholes are virtual anti-particles.[33]

The spin "up" or "down" could be associated with the direction of rotation of the vortex of the holes. Perhaps in one direction (clockwise) the spin is "up," and in the other direction (counterclockwise) the spin is "down". The pulsation rate for the black/white hole exchange is assumed to be in the region of Planck's time. This may have implications for our psychological conception regarding the rate of flow of time. Suppose the values for the Planck scales (mass, time, and length) were changed by assuming a different value for the velocity of light. The evaporation time for a black hole is inversely proportional to the velocity of light. Therefore, if the velocity for light were to increase, the evaporation time for a black hole would decrease. That is, the Planck time, or the basic unit of time in a reality with a higher light velocity, would decrease from our 5.3×10^{-44} seconds. In a similar manner, the Planck length would decrease from 1.6×10^{-33} cm, and the Planck mass would rise from 2.2×10^{-5} grams. For us sentient beings, it would be as if the second and the centimeter contracted. We would experience more units of time in a given subjective interval and more units of length in our subjective space. Time would seem to speed up, and space would expand. An analogy is a film in which each frame is projected more rapidly so that time to the viewer seems to go faster. On the other hand, mass (energy) would become more subtle.

An alternate way of viewing the change in the Planck time can be seen in the following analogy. The repetition time for a periodic wave is determined by the wavelength, or frequency. This can be compared to a clock that clicks as each second passes. A clock can be speeded up by increasing the frequency of the clicks or slowed down by decreasing the frequency of the clicks. Each click determines the length of the second. The clock with a lower frequency experiences a time dilation with respect to the one with a higher frequency. So as the Planck time decreases, this is equivalent to an increase in the frequency of the clicks; time would appear to move faster.

We have noted that a black hole is an antiparticle and a white hole a particle. That is, they have opposite charges. How does this take place? Hawking says that black holes of macro size create and emit particles; by Dirac's equation, these are virtual particle-antiparticle pairs. If the black hole has a net charge, it will tend to repel the part of the pair that has a like charge and attract the opposite. The like-charge particle escapes, and the opposite-charge particle is absorbed. The result is a rapid loss of charge for the black hole. If this process can be assumed on the subquantum level, the black hole and antimatter phase will rapidly lose its charge. The white hole phase will exhibit the opposite charge. So cycling of the black/white hole is also a cycling of the charge from negative to positive and back again. The same is true of angular momentum. The white hole will emit its angular momentum and then regain it in the opposite direction on its return.

The idea of going into the past or into the future is difficult to imagine. Assuming each event is made up of a series of frames, however, an analogy can be constructed. If it were possible to capture the motion of an electron on film, its motion would be made up of a series of frames, each representing a moment in time. Let us suppose that each time the black/white hole goes through one cycle, the previous frame is lost in the black hole and a new frame is brought up through the white hole. The previous frame becomes history and goes back in time, and the new frame becomes the electron's present moment and moves into the future. The black hole seems to remove entropy from our universe, and the white hole supplies us with new information, or negative entropy.

This is reminiscent of the Bondi, Gold, and Hoyle steady state theory, which postulates the continuous creation of matter in every block of space. This theory demonstrates that the density of matter remains constant in an expanding universe, insuring the same look to all observers at all places and at all times. A creation field exists that supplies the universe with negative entropy and thereby affects the anticipated heat death.

It should be reiterated that present instrumentation does not "see" below approximately 10^{-15} cm. Even if new accelerators push this limit somewhat farther, nothing will approach the region of 1.6×10^{-33} cm — the region we have been discussing. Bohm suggests (as does Seth) that it is precisely in this area of the Planck length, where space-time breaks down, that evidence for the implicate order can be found. Experimental probes into this region, however, are probably far off in the future.

Peat's suggestion that the black/white hole may be the gateway to another world of even greater richness and subtlety is echoed by Jungian analyst Marie-Louise von Franz. She states that there are numerous examples of a "threshold" that separates life and death. Von Franz considered the dream of an analysand of hers as representing such a threshold. The analysand did not know Carl Jung personally but was familiar with his work and admired him. The night after Jung's death, of which the analysand was unaware, she had this dream, described by von Franz:

> She was at a garden party where many people were standing around on a lawn. Jung was among them. He was wearing a strange outfit: in front his jacket and trousers were bright green, in the back they were black. Then she saw a black wall which had a hole cut out of it in exactly the same shape as Jung's stature. Jung suddenly stepped

into this hole, and now all that one could see was a complete black surface, although everyone knew that he was still there. Then the dreamer looked at herself and discovered that she, too, was wearing such clothes, green in front and black behind.[34]

Von Franz interpreted the symbolism as analogous to the black hole in physics: the dead Jung disappeared behind an event horizon into another reality. She goes on to relate an old Zulu woman's description of life and death:

> She held out her hand with the palm turned upward and said, "That's how we live." She then turned her hand over with the palm downward and said, "That's how the ancestors live."

CREATING REALITY

We can now relate physics concepts to Seth's description of how reality is constructed. Framework 2 exists in the moment point. It contains a vast unpredictable bank of actions available to all consciousness, from that of a human being to that of an electron. Each consciousness selects actions according to its own inclinations, its own idea of significance, within the rules of the level of consciousness making the selection.

If we again turn to the analogy of picture frames on a reel of film, Framework 2 contains all probable frames to create all possible events. If an electron chooses to go from A to B, it chooses, through its protointelligence, the appropriate frames from Framework 2. The selection process might be visualized as EE units being emitted by the consciousness involved and activating the appropriate subordinate points. Through active information, the EE units form the energy concentration of the subordinate points. The EE units are then slowed down from their normal faster-than-light activities by creating a white hole at the point of electron formation. The EE units carry the energized information (which can be labeled as a future frame) needed to create an electron at point A in space. The frame brought in through the expulsion of energy from the white hole remains in our three-dimensional space an almost infinitely small amount of time.

The white hole turns into a black hole, and the frame is removed;

therefore, there is no permanent or semipermanent electron. Actually, the electron changes every instant; it is continuously created. Old frames leave or go back in time, and new frames come in and go forward in time. There is no indivisible, rigid, or identifiable particle. There is only a pattern created outside of space-time. Any perceived "motion" of the electron is actually a series of projections and injections, which defines a trajectory through three-dimensional space. What we have is a succession of present moments from A to B. Seth comments, regarding this principle:

> Consider this analogy. For one instant your consciousness is "alive," focused in physical reality. Now for the next instant it is focused somewhere else entirely, in a different system of reality. It is unalive, or "dead" to your way of thinking. The next instant it is "alive" again, focused in your reality, but you are not aware of the intervening instant of unaliveness. Your sense of continuity therefore is built-up entirely on every other pulsation of consciousness. . . .

> Remember this is an analogy, so that the word "instant" should not be taken too literally. There is, then, what we can call an underside of consciousness. Now, in the same way, atoms and molecules exist so that they are "dead," or inactive within your system, then alive or active, but you cannot perceive the instant in which they do not exist. Since your bodies and your entire physical universe are composed of atoms and molecules, then I am telling you that the entire structure exists in the same manner. It flickers off and on, in other words, and in a certain rhythm, as, say, the rhythm of breath.[35]

The whole universe flickers on and off like a firefly beneath our level of perception. During the dead or inactive stage, our world is made up of electrons in the black hole stage, the positron stage, or the antimatter stage. According to Seth we do not perceive antimatter. Using his analogy of the wagon wheel, the hub of the wheel represents each successive moment. The spokes at each moment represent the probabilities from which choices can be made. As noted previously, the spokes not chosen can represent a past, since both past and future are chosen in the present moment.

Space as we understand it is made up of mini-black/white-holes that do not have pattern and where matter does not form. In Bohm's terms, space is a vast ocean of energy that is unmanifest or untapped — the implicate order aspect of our three-dimensional world. Matter is manifest, the result of projection and injection that has achieved relative stability. EE units are identified with our spectrum of light when their velocity is slowed down to 300,000 km/sec. In effect, as Bohm puts it, matter is condensed or frozen light. Matter is the condensation of light into patterns.

To summarize, reality is produced by a mixture of thoughts or information coming into contact with conscious but unformed material, what the physicist calls "quantumstuff." The blueprints for reality construction exist within the mind; the patterns are in the implicate order. If the inner blueprint is strong enough, it molds the inactive conscious material into objective reality. While the blueprint exists in the implicate order, the engineer or draftsman is in the superimplicate order, or Framework 3.

The whole process is analogous to the creation of a painting. An artist is given all the tools and materials needed (canvas, paints, brushes, etc.). With them she produces something that she terms objective reality, the painting, but which originated in the creativity of her mind. The materials, which originally were not organized, are now arranged in terms of form and color. In this way it is thought (or mental imagery) that casts itself on unformed space to produce our conventional objective reality.

If the mind is the counterpart to Bohm's significance, the brain is soma. They are rather like two sides of the same coin. The significance side creates the objective world, including the brain. The brain, in turn, interprets the world created by the mind. The two work in a hand/glove relationship; their activities cannot be separated. It is through the brain that the mind manipulates the display. The brain is necessary for us to understand physical existence.

When discussing the creation of reality we must be careful not to fall into the trap of dualism. Actually, mind is forming mind; the unformed field of consciousness is also mind, but of an unformed gestalt. The molding or projection we call matter is a gestalt of conscious elements. The holomovement is one; there is no other. The human mind does not directly create the electron. Rather it results from a seemingly infinite cooperative effort of all minds, from the protointelligence of the electron, to the human mind, to the mind of Gaia, on to the mind of the universe. Our mind's creations mesh perfectly with all others' to produce a common objective reality. As Seth says,

> Each entity perceives only his own constructions on a physical level. Because all constructions are more or less faithful reproductions in matter of the same basic ideas (since all individuals are, generally speaking, on the same level in this plane), then they agree sufficiently in space, time and degree so that the world of appearances has coherence and relative predictability.[36]

While each reality is individualized, all realities are similar enough to allow for the amazing common interplay we all experience. Perhaps that is what the native American sorcerer don Juan meant by "membership in an agreement":[37] we all cooperate to produce a consensus reality.

6

The Mind-Body
Problem from a
New Perspective

*T*he relationship between mind and body was a controversial subject at least as far back as the days of Plato and Democritus. Descartes set the stage for the modern view by assigning matter the attribute of an extended substance (i.e., having a spatial existence), while regarding mind as nonextended. The exact relationship between the two was not clearly defined. As the mechanistic approach moved into ascendancy, the role of the mind gradually diminished until it finally was seen as an epiphenomenon of the brain. But with the advent of quantum theory, the mind-body problem again assumed prominence.

In classical mechanics, if the necessary data of an initial state are known, then the subsequent state can, at least in theory, be determined. In quantum mechanics, the situation is quite different: the initial state can develop many potential subsequent states. As long as a particular resolution is not demanded, the initial state develops deterministically according to Schrödinger's wave equation. However, since physicists assume that only one state can actually occur, this one is selected through observation. Before observation, all possible conditions have a given probability; after observation, this uncertainty is removed, resulting in one choice. The observation is called the collapse of the state vector.

Since this collapse cannot take place without an observer, physicists are now required to consider the role of consciousness in the process. A fundamental assumption is that scientists study nature as dispassionate observers, so it would hardly do to introduce consciousness into the equations of physics. Certain dynamic variables pertaining to self-awareness would have to be represented and quantified, which is clearly not possible at this time. We are left with a quantity called consciousness that appears to be involved in physics, but is nonphysical and nonmeasurable. However, by drawing on Bohm's soma-significance concept and applying a few ideas put forth by Seth, a heuristic argument can be made that may shed some light on this problem.

To help us more fully understand the mind-body issue, we will review some of the concepts of Bohm, Wilber, and Seth that have already been presented. According to Seth, our three-dimensional world is essentially made of consciousness. That is, from consciousness emanate electromagnetic energy (EE) units, which exist just below the range of physical matter. With the help of energy from subordinate points, the EE units are projected into the forms of material objects. Matter, Seth says, is a symbolic construction created by consciousness, out of consciousness. In short, the electron, and all matter, is conscious. Seth points out that although the consciousness of an elementary particle is different from that of a sentient being, it has its own creativity and propensities.

Bohm frequently observes that in some sense nature is alive. He assigns a protointelligence to the elementary particle, as he states in the following passage about his causal interpretation:

> [It] is similar to [Eugene] Wigner's [view] in that it gives the "mindlike" quality of active information a primary role, yet different from it, in that it does not imply that the *human* mind can significantly affect the electron in an actual physical measurement.[1]

Nick Herbert comes to a similar conclusion when he says, "All quantum systems are conscious. 'We' are particularly extensive quantum systems in the brains of certain erect mammals."[2] On this topic Seth says:

> Molecules and atoms and even smaller particles have a condensed consciousness. They form into cells and form an individual cellular consciousness. This combination results in a consciousness that is

capable of much more experience and fulfillment than would be possible for the isolated atom or molecule alone. This goes on ad infinitum . . . to form the physical body mechanism. Even the lowest particle retains its individuality and [through this cooperation] its abilities are multiplied a millionfold.[3]

While the particle is a quantum system and is conscious, it is not an "extensive" quantum system and is not alive as a human being is. Nevertheless, the complexity of the particle is such that it is capable of responding in creative ways to the information in the quantum potential. Its capabilities are analogous to those of a shipboard computer that responds to the radio signal sent out to guide the ship. To Bohm, every level is organized by a more subtle form of significance so that each level, including the level of the particle, has aspects that are both somatic and significant.

Perhaps another way of viewing it is that *all* gestalts of matter are actually manifest portions of quantum matter fields. Matter is an energy state of space itself, or of the implicate order, with atoms, molecules, and so forth representing increasingly complex systems of quantized fields. All matter, biological or not, is a manifestation of its own field. The implications of this are startling. Not only is an electron an explication of its own field, but a human cell can be viewed in a similar manner and, in principle, so can human beings. Our whole body is a manifestation of its own matter field.

Bohm, Wilber, and Seth all describe the hierarchy of consciousness, using different terminology. To Bohm, the spectrum consists of level upon level of consciousness, each of which exhibits material and mental aspects. Each level of soma is organized by a more subtle form of significance, which itself is organized by an even more subtle form. All this tails off into an infinite holomovement. To Wilber, each level has a deep structure and a surface structure, with levels rising through innumerable steps to the Absolute. The lower level is the most dense and fragmentary, the higher levels less dense and more wholistic. Each level has a more limited and controlled degree of consciousness than the one above. The self, as it climbs the ladder of consciousness, can organize and operate the level below. Seth also talks about levels of consciousness, called Frameworks 1, 2, 3, 4, and beyond. Framework 3 programs Framework 2 to create Framework 1.

Another concept of particular interest to both Bohm and Seth is that of motion, which is at the heart of the mind-body problem. In the mechanistic view, matter is seen as inert particles that move through empty space under the action of specified forces. Bohm suggests that this type of motion is illusory. Instead, he sees motion as an enfolding and unfolding process. Each point in space creates matter by this process, and motion can be interpreted as a trajectory of these infinitesimal points enfolding and unfolding from a common implicate order. This is analogous to the track of a particle in a cloud chamber. However, a better metaphor might be the familiar motion picture film. Each frame can be likened to a point of projection and injection. When the film is placed in a projector and run at a rapid speed, each frame dissolves into the illusion of objects moving through space. Wilber refers to this type of enfolding and unfolding as microgeny. Seth calls the same process pulsations.

At least in the three-dimensional universe, we must reject the notion that mind is a force somehow capable of pushing and pulling matter to its bidding. Bohm's view seems more acceptable: the idea that active information guides the motion of matter in a process similar to the quantum potential and superquantum potential. Each level of consciousness is seen as having a somatic aspect, such as a computer, and a significance aspect, such as a computer program. The nonextended substance guides the extended substance as a program guides a computer.

Each level, through its own program, selects from a field of probabilities to determine its next action. A simplified version of this selection process can be seen in the Necker cube, the drawing of an open square box which can be interpreted as either the front of the box seen from below or the back of the box seen from above, one or the other version being chosen by the viewer at any given time. While the probabilities here are only two, the same approach can be used with any number of probabilities. The probabilities are determined by inherent rules governing that level. A large number of choices within the constraints imposed by these rules is important because it allows an entity to be innovative in adapting to changing conditions. Each level can be programmed in innumerable ways within the given constraints; and, as Bohm, Wilber, and Seth all agree, each level is interpenetrated by and is part of every other level.

The interplay of the various consciousness levels of the human body is noted by David Peat. He indicates that each level, while having its own degree of meaning (consciousness), is also conditioned by the needs of the other levels. That is, within the context of the whole, each level uses its own concepts and relationships. In Peats' words,

> The human body is composed of cells which depend upon the whole organism. So at one level the operations of the body can be explained in terms of its constituents, yet at another, these constituent parts must be defined in terms of the goals, operations, and meaning of the whole. In such systems, it becomes clear that at each new level of organization and each new scale of size and complexity, different meanings are present and new and possibly unexpected behaviors and structures manifest themselves. Each level of structure therefore requires its own level of description which contains concepts and relationships that were not present in what went before. But in turn, this level of description is conditioned by what lies above and below in the scale of size and complexity. Living systems, therefore, depend upon a series of levels of descriptions, with none being more fundamental than the other.[4]

Peat goes on to question whether this interweaving of levels is also applicable to matter. He turns to the work of Ilya Prigogine for the answer. Prigogine shows that there is no fundamental level, so reductionism cannot be employed. Whenever a level is considered fundamental, it is found to be dependent on other levels for a definition of its concepts. An excellent example of this is the relationship of the quantum level to the macro level as found in Niels Bohr's Copenhagen interpretation. The quantum level is defined by a macro measuring instrument, which in turn is subject to quantum level laws.

Some of the above ideas are similar to the concept of a "society of minds" proposed by artificial intelligence pundit Marvin Minsky. Minsky suggests that the purposeful operations of the human brain emerge from random interactions of individual or groups of neurons on more elementary levels. (Keep in mind Bohm's idea that order is enfolded in random systems and emerges when the context is broadened; lower levels do not have to be programmed in any detailed way since each level has its own capabilities and equipment necessary to perform the required tasks.) The important ingredient in Minsky's concept is a communication network that unites the various levels so that the wishes and needs of the total brain are carried out in a cooperative manner. Each level is a "mind," specialized and yet creative within the rules of its own position in the hierarchy. Human action, then, is the result of a coordinated effort of all levels, or minds, acting in a purposeful way — hence the expression "society of minds."

Bohm's spectrum of orders reinforces this inclusive view. Based on Bohm's concept of wholeness we could say that any system in physics is dependent on the supersystem in which it operates. Furthermore, the supersystem is dependent on the super-supersystem and so on, to the unknown totality of All That Is. Each level makes a contribution to the total description. We must keep in mind, then, when treating the mind-body problem, that the human being is a subset of an infinite array of more inclusive systems. However, in order to simplify the problem and still describe the concept of the mind-body, we will ignore the supersystems in which the individual is embedded.

Let us suppose that a person wills her right arm to rise. The ego programs, or informs, its soma aspect (brain) accordingly. This programming is not accomplished by the ego moving electrons within neurons. Instead, the brain is guided by its own superquantum potential (ego), which is outside the three-dimensional universe. In an instant (remember that the subjacent universe is not time-dependent) the act of will is known to all levels of consciousness, including the consciousness of the heart, lungs, and other organs, as well as the cells of all the organs and the molecules of all cells down to each individual particle. Each somatic level is aware of constraints on its programmer in making selections from its own field of probability (matter field) in accordance with the requirements of the ego. Each level of the material body uses its own inclinations, its own abilities, to select in conformity with the will of the ego. All the physiological changes necessary are performed in one whole pattern to accomplish the task. It is as if the ego were a composer of a symphonic piece in which each instrument is to be played in a coordinated manner. Each musician conforms to the composition but selects its own variations on the composer's themes. There is freedom of choice at each level, but this choice is constrained or conditioned by the overall pattern instituted by the ego.

Each level, although conforming to the purpose prescribed, is free to be innovative and creative in overcoming possible obstacles. An analogous situation was described by Bohm in discussing his concept of the quantum potential. "With the quantum potential . . . the whole has an independent and prior significance such that, indeed, the whole may be said to organize the activities of the parts."[5] Seth comments on this subject:

> You stand amid a constant vital commotion, a gestalt of aware energy, and you are yourselves physically composed of conscious cells that carry within themselves the realization of their own identity, that cooperate *willingly* to form the corporeal structure that is your physical body.[6]

Psychologist Donald Campbell's concept of downward causation is similar. Campbell says that "all processes at the lower levels of a hierarchy are restrained by and act in conformity to the laws of the higher levels."[7]

As each musician enjoys freedom of expression within certain parameters, each level also exhibits that combination of constraint and choice. This is not a deterministic push-pull system; it is a correlated effort of free gestalts of consciousness. So, in a similar manner, with the thought, "Raise my right arm," a metaphorical symphony is composed, parts are arranged for each instrument (or levels of consciousness), and the total orchestra performs the work.

To return to the metaphor of the motion picture, each note in the symphony, or each infinitesimal movement of the arm is analogous to a single frame on the reel of film. As in a film, no one notices that the trajectory of the arm is not a continuous movement through space but a track of a near-infinite number of projections from the implicate order. The entire production is arranged and produced behind the scenes of the three-dimensional universe through the interaction of layer upon layer of consciousness, each guiding its level through the proper motions. There is no need to push matter through space, nor is there a need to bring human consciousness into physics, unless physics moves into the superimplicate order. Similarly, there is no need to bring onto the stage the composer of the symphonic work. The performance can be fully experienced with the musicians playing their parts of the musical score under the direction of a conductor, the conscious mind.

Seth discusses the body's cooperative effort:

> In the entire gestalt from cellular to "self" consciousness, there is a vast field of knowledge — much of it now "unconsciously" available — used to maintain the body's integrity in space and time. With the conscious mind as director, there is no reason why much of this knowledge cannot become normally and naturally available.[8]

And:

> All the cells in the body have a separate consciousness. There is a conscious cooperation between the cells in all the organs, and between the organs themselves. . . .[9]

And:

> "You" are the one who decides to walk across the floor, and then all of these inner calculations take place to help you achieve your goal. The conscious intent, therefore, activates the inner mechanisms and changes the behavior of the cells and their components.[10]

Neuroscientist Candace Pert has a similar interpretation, as revealed in *Brain / Mind Bulletin*:

> Pert discussed her theory on how the AIDS virus gains entry to cells. She and her husband, Michael Ruff, talked about how their NIMH peptide studies led them to think of consciousness as being "located in every cell of the body."[11]

By redefining motion as a series of projections and injections, and by viewing consciousness as information rather than a force, the mind-body problem may be resolved. Consciousness need not be brought into the physical world in the manner of an electromagnetic field or a gravitational field. If such an attempt were made, consciousness would have to be defined as a mass or a charge. That is precisely the problem with the approach taken by Wigner. He sees the mind as a separate entity which, in some unknown fashion, interacts with another entity called matter. In effect, the mind somehow enters time and space, and within the present laws of physics purposefully affects the motion of matter. John Stewart Bell points out the difficulty in developing such a theory.

> As soon as you try to put such theories down in mathematical equations, as soon as you try to make them Lorentz invariant[12], you get into great difficulties. For example, the interaction between the mind and the rest of the world, how does that occur? Does that occur over a finite region of space, at an instant of time? Clearly not, because that is not a Lorentz invariant concept.

> And the only way to get such a consistent description, if you assume the mind has access to a single point in time, is to also assume it has access to a single point in space.[13]

Using our interpretation, including that of the transcendent order, the mind does not have to be Lorentz-invariant. The programming is not done in the three-dimensional world but in the preferred frame of reference of the holomovement. In Bohm's terminology, the mind does not influence matter in a direct causal way within the explicate order; the connection is through resonances in the implicate order. These resonances, or nonlocal correlations, are not the result of faster-than-light signals within space-time. All space — and, for that matter, all time — is enfolded in the holomovement, and events in the implicate order are not ordered sequentially. Mind, then, seems to be some kind of field not in our three-dimensional world, not manifest locally, but everywhere, transcending our world.

If physicists remain concerned only with the explicate order, then they will probably not be able to treat consciousness within the physics discipline. However, if the study of physics is broadened to include the implicate and superimplicate orders — the levels of choice, programming, Herbert's choreographer — then mind becomes an object of study.

Our view of the mind-body problem stands in direct opposition to the classical approach. In classical physics, the elementary particle was acted upon by blind forces which, in aggregate, create our macro world. The causal arrow was from the elementary to the complex. The new physics reverses the process: causation points downward from the macro to the micro. The implicate order is primary; the particle is secondary. There is a hierarchy of hardware-software levels, the lower levels being defined by the higher-level software. There are programs within programs, the higher levels including the lower levels and encompassing more than the lower levels. The whole is greater than the sum of its parts, but the whole does not push the elementary particle around in the Newtonian fashion. Instead, guidelines are set by the whole, and the normal explication of resonances of the various levels moves in concert with the overall program.

7

The

Measurement Problem:

Schrödinger's Cat

Revisited

---

*T*he quantum measurement problem exemplifies the bizarre nature of the subatomic world from the perspectives of our everyday reality and classical physics. Throughout the text, we have discussed various aspects of the measurement difficulty as it applies to the subject under discussion. In this section, we will discuss its ramifications in more detail and then review Bohm's and Seth's solutions to this puzzle.

In 1924, Louis de Broglie suggested that each portion of matter has a wave associated with it. This was subsequently confirmed in the laboratory, thus uniting particles and fields, at least conceptually. At first glance the idea that particles have both particulate and wave aspects is not so odd. There are many waves in nature that result from swarms of particles moving in a particular way. In a water wave, for example, each bit of water acts like a particle, and the particles in aggregate function like a wave. But if the water is in a container, and you remove enough of the water particles from the container, the wave aspect disappears. Not so with quantum particles: when a single quantum particle is observed, it shows up as a particle, but when it is unobserved, it seems to exist as a wave.

This peculiar feature of quantum particles is demonstated in an especially striking manner in the electronic version of the two-slit experiment (see pp. 134-35). If a large number of particles pass through the slits, an interference pattern builds up on the detector panel behind the slits, indicating that a wavelike disturbance passed through both slits simultaneously. Contrary to expectations, though, if the particles pass one at a time through the slits, an interference pattern of waves still results, after a long enough time and a large enough number of single particles. Physicist John Barrow described the two-slit experiment using neutrons:

> If we fire the neutrons slowly, one at a time, towards the screen so that we can watch the film developing neutron by neutron, and so avoid any obvious interaction between different neutrons which would lead to interference, then we still find the interference pattern being built up bit by bit. More striking still, we could set up many identical versions of this experiment all over the world and fire just one neutron towards the slits in each of them at a prearranged moment. If we add together the results from all these completely different experiments we would find that the net result would look like the wave interference pattern! The single neutrons seem to be able to interfere with themselves. This is indeed "one hand clapping."[1]

Barrow is saying is that a single neutron, when not being observed, shows the properties of a wave.

Schrödinger (among others) placed this strange situation on a mathematical footing by developing a wave equation. Initially, he believed that his equation would apply within the context of classical physics, in a manner similar to the equation for electromagnetic waves. But this belief was short-lived. The wave as defined by Schrödinger is not directly observed. It cannot be detected because it does not carry energy. If it does not carry energy, an interpretation problem arises. Ordinarily, the energy of a wave is proportional to the square of its amplitude. Without energy content, what does the amplitude of this wave signify? Max Born came up with an answer. The amplitude squared in this case does not indicate the wave's energy but rather its probability. Since the wave is associated with a particle, the probability gives physicists a measure of the likelihood of finding a discrete energy packet, or particle, at the point associated with a given amplitude.

Schrödinger's wave equation thus became a calculating device rather than describing a "real" wave in the classical sense. Using the equation in this way leads directly to the measurement problem. A short digression on how calculations are performed will clarify the philosophical dilemma.

In a simple case, suppose the physicist wishes to study the state of an electron at a particular moment. Schrödinger's equation is applied, and a solution is found, called the wave function. The wave function indicates how the electron moves around space in a given amount of time. For example, suppose an electron shoots out of an atom. This event is represented mathematically by a spherical wavefront moving outward in ever-widening circles from the atom. Depending on the attribute the physicist wishes to measure, the wave function is decomposed into a given waveform family.

The decomposition is accomplished by a procedure called Fourier analysis in which any periodic wave is expressed as a spectrum of sine waves. (With the aid of computers Fourier's procedure has now been greatly extended so that a periodic wave can be reduced to waveforms other than sine waves.) Schrödinger was able to show us that each dynamic attribute can be associated with a particular spectrum of waveforms. As examples, an impulse wave is associated with the attribute of position and a sine wave with the attribute of momentum. Each element in the spectrum represents a value for the attribute the physicist wishes to measure.

Before a measurement is made, the electron could be anywhere in the universe at any given time. After a measurement is made, all the terms in the infinite series go to zero, except the one designating the point of measurement. The wave function collapses to a point instantaneously. It is easy to see why most physicists disregard pictorial explanations of this occurrence and use Schrödinger's equation simply as a mathematical algorithm. But to some physicists, this approach is philosophically unsatisfactory. The conflict about how to view the instantaneous wave collapse indicates the crucial importance of the act of measurement in quantum physics.

Before we look deeper into the concept of measurement, let us take stock. If we forget about the measured electron for a moment, we could say that the electron is a wave. The problem arises when we *observe* the electron — and it suddenly becomes a particle. We could then assume that the electron is a particle, but this assumption clashes with the experimental facts of the two-slit experiment. How can a single particle interfere with itself? The only explanation seems to be that the electron is a wave when we are *not* looking and a particle when we *are* looking. When the electron is a wave, it is described by a wave function. The wave function is decomposed into a series of different states, each with its own probability for existence. In other words, when we are not looking, the electron seems to occupy a state of *potentia*, a world where all probabilities exist side by side. The only thing differentiating these probable states is their likelihood of coming into existence.

This apparent need for an act of observation, or measurement, to bring an electron into existence has been explained by physicists in several ways. The most widely accepted explanation is the Copenhagen interpretation of Niels Bohr. Bohr divided reality into two parts, the quantum world and the familiar world of classical physics. When the electron occupies the quantum realm, it acts like a wave and exists in a state of *potentia*; we cannot directly perceive this state. Bohr's interpretation assumes the reality of the classical world and turns to quantum theory to describe the contact of the measuring instrument with the unseen world. In other words, the starting point is the classical world with its instruments, rather than the elementary particles themselves. According to the Copenhagen view, our link with the quantum world is through a measuring device.

In some inexplicable way, the measuring instrument converts the counterintuitive quantum world into our familiar physical world. The instrument does not necessarily have to be in the laboratory. For example, our visual system (eye, optic nerve, brain, etc.) can be considered a macro measuring instrument. Exactly how the conversion is made from the quantum world to the macro world was not addressed by Bohr. It was not until 1932 that John von Neumann attempted a more complete explanation of the quantum measurement problem by bringing in the observer as an essential element.

What von Neumann attempted to discover was the site of the wave function collapse. At what point does this mysterious conversion take place? Exactly when does the electron go from the ghost world of the quantum level into our everyday macro reality? Von Neumann attacked the question by visualizing the measurement problem as a series of steps, or a chain with a large number of links, through which he could determine the point of the collapse.

But a problem quickly became evident: the measuring instrument itself is composed of quantum objects; it also can be described by quantum theory. And, as Bohr pointed out, if you wish to treat the measuring instrument as a quantum device, then a second measuring instrument is needed to bring the first one into reality. As one can see, this leads to an infinite regression. Von Neumann, following his own logical steps, concluded that any place on the chain could be the site of the collapse, which put him back to square one. If there is no indisputable place where the unseen quantum world turns into our macro world, von Neumann reluctantly concluded, then the collapse takes place in the physicist's mind. That is, human consciousness is needed to collapse the wave function and thereby create reality. So von Neumann stretched the Copenhagen view by adding an observer as well as a measuring device to the puzzling paradox.

Even if we accept this interpretation (and many physicists do not), the problem is not resolved. How do we define an observer? Must the observer be human? Can the eye and mind of a frog collapse a wave function? Can an amoeba, or even an electron, do the job? If this were to be the case, literally all gestalts of matter would need to possess some version of a mind.

This brings us to the work of David Bohm, who sees the measurement problem in terms of the implicate and superimplicate orders. In the 1920s, de Broglie introduced the concept of a pilot wave, which is associated with the particle and, in effect, tells the particle how to move. In developing this idea de Broglie encountered many mathematical problems, and the followers of the Copenhagen school completely rejected his approach. But Bohm revived the concept in his causal interpretation and placed it on a sound mathematical footing. He proposed a real particle guided by a wave and described by a transformed Schrödinger equation. The equation was modified to contain an additional term Bohm labeled the quantum potential. Bohm's wave was an information wave, and its effects depended on its form, not its magnitude. The resulting quantum potential could be activated anywhere; its effects did not diminish with distance. Using this theory with the two-slit experiment, the quantum potential is seen as guiding each particle so that an interference pattern forms on the detector panel.

The faster-than-light aspects of the quantum potential led to the concept of nonlocality. Nonlocality is required by Bell's theorem if the physicist wishes to discuss objective particles. With nonlocality, faster-than-light effects are a must, but this, as we have noted, violates Einstein's special theory of relativity. Bohm overcame the problem by assuming that the realm underlying the world of objective phenomena is an undivided, seamless whole. He reached this formulation by applying his causal interpretation to the quantum field. The underlying whole he identified with the holomovement, an infinite spectrum of implicate orders engaged in a process of enfolding and unfolding of wave packets from the main body. All wave packets are interconnected by the seamless whole from which they spring. Furthermore, the passive implicate order is organized by a superimplicate order. It is the superimplicate order that determines the "real" attributes of the particle.

The measurement problem, from this perspective, becomes a function of the information field. According to Bohm's conception, the implicate order is a passive information field potentially active everywhere. In our example of the electron being ejected from an atom, we know that the particle has a probability of being at any number of positions (A, B, C, . . . etc.). According to the Copenhagen view, if a measuring instrument is placed at point B and a particle is found, the wave aspect of the particle collapses to point B. That is, the probabilities at A, C, and all other positions become zero. Rather than interpreting this as a collapsing wave function, Bohm would say that the implicate order was activated at point B by the superimplicate order, or superquantum potential. The information field at other points was not activated, and hence no particle was explicated at any point except B. From our three-dimensional perspective, all points except B are still in a random state; no explication has taken place.

When Bohm says that the superquantum potential explicated the electron, he means that the "mind" of the electron created its own explication. Consciousness need not be sentient to create a particle. In some sense, the particle itself is alive and is capable of activating the implicate order. So instead of a wave collapsing to a point to produce a three-dimensional particle, Bohm says that the superquantum potential activates information at the point of explication, and the passive implicate order, in turn, creates a particle by the enfolding and unfolding process. The idea of a collapsing wave function is not necessary.

In Bohm's approach (and Seth's), the electron is responsible for its own creation. When a sentient being sees a chair, how do all the particles know to explicate themselves so that a chair is created? Before the details of this process are discussed, another concept is needed, best illustrated by the thought experiment called "Schrödinger's cat."

Schrödinger, one of the founders of quantum mechanics, wondered about the indeterminacy displayed on the quantum level and the apparent lack of it in the macroscopic world. As an example, let us suppose that a quantum system can evolve in two possible ways, represented by two waves superimposed. The two waves interfere with each other to produce a complex pattern. At this juncture, a measurement is made, and one of the wave patterns is selected. Keep in mind that, according to the Copenhagen interpretation, the measurement is made by coupling the macro measuring devices to the quantum system. When the "answer" is obtained (one possibility selected), an irreversible process takes place. The complex wave pattern is instantaneously destroyed, and one wave pattern is left. This whole process takes place because we enlarged the quantum system to include macro measuring devices.

As we have seen above, if the whole universe is quantum-mechanically described, the wave function can be collapsed only by the mind. Without the mind, the universe exists in states of *potentia*. That is precisely the paradox that Schrödinger wished to illustrate with his famous thought experiment, which placed the quantum paradox directly in our macro world.

The thought experiment involves the following situation. A live cat is placed in a sealed box. Along with the cat is placed a device subject to the decay of a radioactive substance. After one hour, there is a 50% chance that one atom will decay. If the atom decays, the device is tripped, a container of poison gas is shattered, and the cat is killed. The experiment takes place in a laboratory, which includes a physicist who is handling the experiment. If von Neumann's analysis is correct, the cat is in a superposition of many states; that is, it can be alive or dead or anything in between. Since Schrödinger's equation is linear, solutions are possible that incorporate a cat that is partially alive (say, 75%) and partially dead (25%). The cat is in a state of potential waiting to be realized.

How is the cat saved from this suspended animation? By the physicist opening the box and looking. At that point, and only at that point, does the cat finally find out whether it is alive or dead. According to von Neumann's idea, the same would be true if the cat were replaced by a human being. However, in the case of a person, the physicist could ask the subject how it felt to be in a suspended state (assuming that the subject is alive and available for questioning when the box is opened). We all know what the answer would be: the subject would laugh and attest to having been alive all along. But you and I know that this violates quantum mechanics, a theory that has not been wrong for fifty years! Schrödinger's cat clearly puts the quantum dilemma in our own domain, with consequences detectable on the macro level.

Bohm would interpret the travail of Schrödinger's cat differently. He would assume that the cat is always in a definite state, alive or dead, even before the physicist comes to the rescue. If the physicist finds the cat alive upon opening the box, Bohm would say the cat has always been alive. All those other possible states were never selected for explication by the superquantum potential of the cat. They are still in the implicate order, not manifested in our three-dimensional world. In Bohm's view, the physicist was *not* needed to bring about the living state.

Seth has not, to my knowledge, directly commented on the paradox of Schrödinger's cat. But he has made some comments about our reality that are relevant to this problem. A short digression into some of the mathematical devices used by physicists will help make Seth's views more understandable.

The theories of relativity and quantum mechanics operate with different mathematical concepts of space. Relativity is ensconced in the four dimensions of Riemann space (three dimensions of space and one negative dimension of time), while quantum mechanics operates in the infinite dimensions of Hilbert space (discussed below). Unfortunately, the two theories have not been united, nor does there exist a mathematical formulation of which both these theories could be considered special cases.

Three-dimensional space proved inadequate in quantum theory, since a quantum system must have enough dimensions to handle all independent possibilities. An example is a quantum system composed of two electrons. If it is assumed that each electron occupies the same three-dimensional space, then simple results, such as the conservation of momentum when they collide, cannot be calculated. To do so, three-dimensional space must be assumed for each electron, or a total of six dimensions.[2]

As the number of possibilities goes up or the number of electrons goes up, the number of dimensions increases. In some problems, the possibilities are unbounded, and the physicist uses the infinite-dimensional Hilbert space for calculations.

The concept of Hilbert space originated in the early 1900s with mathematician David Hilbert's theory of infinite-dimensional spaces. We human beings supposedly inhabit a three-dimensional space; that is, any point in space can be represented by an ordered sequence of three coordinates: length, width, and depth. In a similar manner, a point in Hilbert space can be represented by an infinite sequence of numbers. Multidimensional space is not difficult to handle mathematically. Higher-dimensional spaces, called "phase spaces" by physicists, are used to represent a large number of pieces of information. For example, if there are ten pieces of information, the physicist will consign them to a ten-dimensional phase space. If the information is infinite, it can be handled mathematically in the infinite-dimensional Hilbert space. While a space of this kind cannot be visualized, the concept proved invaluable in the theoretical physics of Heisenberg and Schrödinger, who saw quantum particles as patterns in this infinite-dimensional space. Thus, Hilbert space is seen as nothing more than a very useful algorithm for predictive purposes. To Bohm, however, Hilbert space is more than a mathematical tool: it represents the implicate order.

One of the reasons we have difficulty envisioning such multidimensional spaces is because we assume that an objective world exists independent of the observer. In so doing, we define the motion of objects in three dimensions of space and one of time. Rudy Rucker, in *The Fourth Dimension*, suggests an alternative approach. Suppose we assume that reality is basically the perceptions of various observers; that is, our perceptions are primary, not the external objects. Then the description of matter in a three-dimensional space turns out to be just one possible ordering of a limited number of perceptions. We would also need coordinates for such thoughts and impressions as sweet and sour, good and bad, cold and hot, dark and light, and so forth. It is easily seen that with this procedure we would require an endless number of axes. Furthermore, since we each have our own perceptions, we would have to multiply the coordinates by the number of observers. Certainly, some sort of Hilbert space would be required.

For another view of these ideas, we now turn to Seth. If Seth were to consider the problem of Schrödinger's cat, his evaluation would be quite different from that offered by most physicists and philosophers. Seth asserts that all gestalts of consciousness create their own reality. Although some contemporary physicists would agree partially, most would limit this ability to the consciousness of a sentient being. Since Seth views even an electron as conscious, he sees this ability as extending to all matter. Bohm largely agrees; the electron exhibits a protointelligence without self-awareness, but this ability to create its own reality is assigned to the superquantum potential. Seth makes an interesting point that relates to the paradox of Schrödinger's cat. In a session recounted by Jane Roberts, three people were present, Jane, her husband, and one other person, along with the family cat. Jane (while in trance) picked up a glass from a table, and Seth remarked:

> None of you sees the glass that the others see. . . . Each of the three of you creates your own glass, in your own personal perspective. Therefore you have three different physical glasses here, but each one exists in an entirely different space continuum.
>
> Physical objects cannot exist unless they exist in a definite perspective and space continuum. But each individual creates his own space continuum . . . I want to tie this in with the differences you seem to see in one particular object. Each individual actually creates an entirely different object, which his own physical senses then perceive.[3]

The key thought here is that *each one of us creates our own three-dimensional space*. The reason we feel we share the same space is because of our ability to relate to and communicate about the common features of our environments. Our spaces are similar because we are all on the same level of consciousness and have the same neurological equipment. Jane Roberts' cat also creates the glass, but if it could communicate with us, its space would be described differently. By extrapolation, the electron creates its own environment, although it is quite different from ours or the cat's. What, then, is the reality of the glass? If we take each consciousness that can produce and then perceive the glass, the reality of the glass exists in a multidimensional space made up of each space of each consciousness involved. Any number of people, cats, and other beings can create a glass because we are dealing with an infinite-dimensional space, a kind of Hilbert space. So while each one of us seems to exist in only three-dimensions, we are part of a multidimensional space.

Imagine two people playing a game of chess by mail, one in London and the other in New York. Each has a chessboard with the pieces identically arranged. The moves described in the players' correspondence change the arrangement of pieces on each board. While the game is similar to a game played with one board, there are actually two boards in existence. In a similar manner, each of us has our own chessboard or three-dimensional space, but we are all playing the same game. Reality, then, is made up of all the chessboards, or all the three-dimensional spaces, with each of our realities being a part of the total. The communication that makes possible the changes in the game takes place in an underlying telepathic order to which all players make contributions and to which all have access.

Now, back to Schrödinger's cat, using Seth's approach. To actually conduct such an experiment, the physicist would create a three-dimensional space that includes the closed box, the laboratory, and all seen objects. Since the cat inside the box is not perceived, it is not created. When the physicist opens the box, then the cat is created, dead or alive. According to Seth, the cat, too, would create a three-dimensional space, but a different one than the physicist. If the cat, with the cooperation of the physicist and all other consciousness, decides to stay alive while it is in the box, then according to its three-dimensional space, its perspective, it is alive and not in a state of suspended animation. The physicist may perceive the cat as neither dead or alive, but that is because the physicist has not yet brought the cat into the physicist's own three-dimensional reality. Viewed in this way, the paradox disappears.

Everybody's perceptions — the perceptions of all living creatures, of all gestalts of matter, including the electron — together create an infinite-dimensional space. That space may be the true reality for our universe. Consciousness can construct the universe in any number of ways, with no one configuration more true than another. Each has its unique focus, and each is important. All these perceptions together make a common reality.

8

A Closer Look
at Bell's Theorem

*I*n October 1927, the fifth of the Solvay conferences (named for Belgian industrialist Ernest Solvay, who sponsored and funded them) took place in Brussels. All the eminent physicists of the day were in attendance; Einstein, Bohr, de Broglie, Heisenberg, and Born were among the participants. That year Bohr had clarified his thoughts concerning the wave-particle duality and presented a paper on the subject at the conference. His ideas incorporated the now-famous Copenhagen interpretation of quantum mechanics. While the paper was accepted by most physicists, the one man Bohr wanted most to convince remained skeptical. That man was Albert Einstein. Einstein remarked later:

> There is no doubt that quantum mechanics has seized hold of a beautiful element of truth, and that it will be a test stone for any future theoretical basis, in that it must be deducible as a limiting case from that basis, just as electrostatics is deducible from the Maxwell equations of the electromagnetic field or as thermodynamics is deducible from classical mechanics. However, I do not believe that quantum mechanics will be the *starting point* in the search for this basis, just as, vice versa, one could not go from thermodynamics . . . to the foundations of mechanics.[1]

Einstein could not reject quantum theory because it failed to agree with experiment since in that respect the theory proved successful.

But since quantum theory makes only statistical predictions, it must, Einstein felt, by its very nature, omit certain elements of reality; therefore, it was incomplete and would eventually have to be included in a broader, more complete theory.

Thus began the famous Einstein-Bohr debates regarding the validity of quantum mechanics as a conclusive picture of reality. Einstein participated in the debates by presenting a series of thought experiments designed to show that quantum theory left something out and therefore was deficient. Bohr, at least in the minds of most physicists, answered each experiment in a coherent and consistent way. Nevertheless, Einstein was not convinced. In 1935, he released his most famous salvo in an article in *Physical Review*. It was co-authored by two American physicists, Boris Podolsky and Nathan Rosen. The article questioned the validity of the Copenhagen interpretation and posed what Einstein viewed as a paradox.

The authors began by defining what they meant by a complete physical theory. Two assumptions were stated:
1. A complete physical theory must have a counterpart for every element of physical reality.
2. If an element of reality can be defined precisely (probability of one) by a physical quantity which can be obtained without disturbing the system, then that element can be considered real.

(As Bohm pointed out, these assumptions contain an implicit assumption: that the world can be analyzed and described by distinct and separate elements.) Einstein, Podolsky, and Rosen then proposed the following experiment (known as the EPR experiment).

Suppose two particles, A and B, are in close proximity in space. They are moving, and therefore have momentum. Heisenberg's uncertainty principle does not allow the physicist to measure the position *and* momentum of a particle simultaneously without obtaining uncertain results. However, the total momentum of particles A and B and the distance between the particles can be measured with certainty. Now, suppose that the two particles interact and fly away from each other, at near the speed of light, to opposite ends of our galaxy. This would preclude any signal from passing between them because of the restriction of the special theory of relativity, which states that all signals are confined to the speed of light or less. Suppose a physicist is stationed on one side of the galaxy and measures the momentum of particle A. Since the total momentum is the same before and after the interaction (momentum is conserved), the physicist automatically knows the momentum of particle B. If the position of particle A is also measured, the position of particle B can be determined. It is true that in measuring the position of particle A, the physicist disturbed it and therefore made its momentum uncertain; however, this should not disturb the momentum of particle B. (If such did happen, it would mean that a signal passed from A to B in violation of the principle of relativity.) Through this process, the physicist has determined with certainty the momentum *and* position of particle B without disturbing the system of particle B.

The determination made through this thought experiment allowed the physicist to violate one of the basic tenets of quantum theory, the uncertainty principle. Einstein felt that he had thereby achieved his objective and proved that quantum theory was incomplete. Based on his assumptions, particle B had objective existence without being measured. Of course, Einstein assumed local causality as a given. Bohr did not attack the local causality assumption; he considered the flaw in Einstein's approach to be the assertion that B had a real existence without being measured.[2] The two continued to differ on this point. Max Born summed up Einstein's views:

> The generation to which Einstein, Bohr, and I belong was taught that there exists an objective physical world, which unfolds itself according to immutable laws independent of us; we are watching this process as the audience watches a play in a theater. Einstein still believes that this should be the relation between the scientific observer and his subject.[3]

To Einstein, the EPR paradox presented an insoluble problem. He felt that if quantum theory is a complete theory, then the physicist has only two choices.
1. Particle A somehow communicated with particle B at a speed faster than light and thus violated the local causality principle, or,
2. Particle B is not really there until measured.
For Einstein, both choices were equally unacceptable, and he concluded that quantum theory must be incomplete. There the argument

foundered until John Stewart Bell appeared on the scene. Bell's contribution was to question the local causality assumption; that is, particle B cannot be affected by particle A if it requires a signal with a speed faster than that of light.

In 1964, while on leave from CERN[4], Bell published in the *Journal of Physics* what became known as Bell's theorem. He noted that for Einstein to be correct and quantum mechanics to be incomplete, there must be additional (hidden) variables that would restore local causality to the EPR paradox. Bell first refuted a conclusion arrived at by John von Neumann in 1932, which stated that even without a locality requirement, no hidden variable interpretation is possible. Among other evidence, Bell pointed to Bohm's paper of 1952, in which a consistent hidden variable theory was indeed produced in violation of von Neumann's proof. However, Bohm's theory did have a nonlocal feature.

Bell's theorem assumes that a *local* reality exists with particles A and B as described in the EPR experiment. If the locality assumption is true, this leads to what is known as Bell's inequality. If the inequality is violated, then the locality assumption is false, or any reality that underlies the EPR experiment must be nonlocal. In effect, Bell's theorem provides a method for comparing predictions between a local realist approach and that of standard quantum theory. By approaching the problem in this way, Bell provided a way for experimental physicists to enter the argument.

At Bell's suggestion, several experiments were performed that showed Bell's inequality to be violated, but the results left some room for doubt. In 1976, Alain Aspect proposed what was to be the definitive experiment and finally performed it in 1982. This verified the strongly correlated quantum theoretical predictions, violated Bell's inequality, and proved that the world beyond the quantum level is nonlocal.

Nick Herbert sums this up:

> The toughest limitation on a local interaction is how fast it can travel. When you move an object A, you stretch its attached field. This field distorts first near object A, then the field warp moves off to distant regions. Einstein's special theory of relativity restricts the velocity of this field deformation to light speed or below. According to Einstein,

no material object can travel faster than light; not even the less material field warp can travel so fast.

Non-local influences, if they existed, would not be mediated by fields or by anything else. When A connects to B non-locally, nothing crosses the intervening space, hence no amount of interposed matter can shield this interaction.

Non-local influences do not diminish with distance. They are as potent at a million miles as at a millimeter.

Non-local influences act instantaneously. The speed of their transmission is not limited by the velocity of light.

A non-local interaction links up one location with another without crossing space, without decay, and without delay. A non-local interaction is, in short, *unmediated, unmitigated,* and *immediate.*

Despite physicists' traditional rejection of non-local interactions, despite the fact that all known forces are incontestably local, despite Einstein's prohibition against superluminal connections, and despite the fact that no experiment has ever shown a single case of unmediated faster-than-light communication, Bell maintains that the world is filled with innumerable non-local influences. Furthermore these unmediated connections are present not only in rare and exotic circumstances, but underlie all the events of everyday life. Non-local connections are ubiquitous because reality itself is non-local.[5]

Bell's theorem and Aspect's experimental verification provided physicists and philosophers with an invaluable tool and caused some proposed models to be discarded. As an example, computer scientist Ed Fredkin's idea that the universe is a giant computer must be rejected by these criteria. Since the computer (at least as we understand it today) is local, reality cannot be computerlike. Nor can the universe be viewed as a giant brain, for, again, by present standards, the brain is local. Of course, Einstein's view that the world can be disassembled into separately existing elements also must be rejected. The model of billiard balls in space is no longer tenable; the classical metaphor must be supplanted.

The mind, however, can be nonlocal — and that opens up possibilities. One of the phenomena frequently discussed in parapsychology circles is extrasensory perception (ESP), the ability to obtain information from outside the normal range of the senses. While the existence of ESP has never been accepted by the scientific community at large and would be excluded under the classical metaphor, it is not ruled out by the Bell-Aspect criteria. Indeed, these and other strange happenings may be explainable by nonlocal means.

The Bell-Aspect criteria verify Bohm's notion of unbroken wholeness. A quantum system cannot be analyzed by considering independent existent parts that have fixed dynamical relationships. The parts are interconnected on the quantum level; local entities are not possible. According to Bohm's causal interpretation, distant quantum systems have a direct connection through the quantum potential. The systems literally cannot be far enough apart to behave in a completely independent manner. The quantum potential depends on the system as a whole — including the entire universe. When all classical potentials have been reduced to zero, the particles still can be connected by the quantum potential. The important concept is *wholeness*.

The notion of separate entities requires an understanding of what is meant by a signal. According to Bohm and Hiley,

> In the theory of relativity, the concept of a signal plays a basic role in determining what is meant by separability of different regions of space. In general, if two such regions A and B are separate, it is supposed that they can be connected by signals. Conversely, if there is no clear separation of A and B, a signal connecting them could have little or no meaning. So the possibility of signal implies separation, and separation implies the possibility of connection by a signal.[6]

To Bohm the regions A and B are not separate; therefore, the problem of whether a signal is transferred is not relevant. Bohm views relativity as a statistical effect, not as absolute. There seems to be a transfer of information only when we observe the correlations at two separate locations. If, on the other hand, we view only the local area of either A or B, their statistical properties appear to be independent. The faster-than-light aspects do not appear.

Bohm suggests that the EPR experiment can be viewed using spin rather than position and momentum. He supposes an arrangement where pairs of electrons are shot out in opposite directions. The structure is such that if the spin of one electron points up (*U*), the other electron must have a spin pointing down (*D*). If the spin of one electron is measured, say on Earth, the electron on the distant galaxy "knows" that its spin must be the opposite. (For simplicity's sake, we are ignoring the fact that all components of spin cannot be measured at a given time. The principle involved can still be established.) This instant knowing violates special relativity theory, assuming that a signal is passed. But how do we know that a signal is passed? The measurements of the electron on Earth consist of a random series of *U*s and *D*s. If there is a correlation with the electron on the distant galaxy, it too will have a correlated random series. Both the observer on Earth and the observer on the distant galaxy see a random sequence. When this order is changed on Earth, the order on the galaxy would presumably also change, but the observer would not be aware of it because one random situation is the same as another. If the observer on Earth were to signal the observer on the distant galaxy via Bell's theorem, the latter could not pick up the signal. Therefore, no signal is passed. Hiley says the following regarding signal transfer:

> It's not clear to me that such a possibility exists. If we go back to Aspect's experiment, although the quantum potential shows there is an instantaneous connection, when we look at the statistical properties of the particles at each end of the connection, they (the particles) appear to be independent; it's only in the correlations that we see the non-locality. It's not clear to me that those correlations can ever be transformed into a signal which makes things go backwards in time.[7]

In essence, the signals are scrambled beyond retrieval by ordinary three-dimensional means and cannot be received through our everyday reality. Signal transfer must be accomplished in an underlying order of deep reality, which does not have separate parts; it is one undivided whole.

Jack Sarfatti sees no violation of the special theory of relativity in the Bell-Aspect criteria. To quote:

> There is no violation of Einstein's theory of relativity because the information transfer does *not* require the propagation of energetic signals. The quantum information utilizes energy already present at a particular place.[8]

If we turn to Seth, we can approach the problem in another way. To signal is to communicate information. Starting on the level of human consciousness, let us assume that we sentient beings get information in two different ways: through our senses, and from outside our senses. Our faculties of sight, hearing, smell, taste, and touch allow us to perceive stimuli (information) from both inside and outside our bodies. These are our normal forms of perception, and this information can, in principle, be measured with a physical device. Seth describes the process as certain effects of the mind, emotions, and so forth, stimulating an objective (but unformed) reality of subordinate points, which then is manifested into our personal three-dimensional world.

Presumably the brain and nervous system are also formed in this procedure. The brain then interprets this mixture of mind and passive energy, giving us that feeling of a reality being out there. The brain is the physical counterpart of the mind; it is how the mind translates itself into three-dimensional reality in order to interpret its handiwork and manipulate the display. In a sense, the brain allows us to tune in on, or focus on, our created reality in a clearer way. An energetic signal is required to stimulate the brain and nervous system so that perception can take place. This normal sensory information requires explication (using Bohm's term). Explication sequences our experience so that we seem to live in a temporal continuity. Once explicated, the universe follows the three-dimensional rules of the game. Presumably, the special theory of relativity is a subset of those rules. An energetic signal (one created through explication) is limited by the velocity of light in our universe.

But what about other kinds of information? We are all aware of thoughts and images that do not come from the external world of our sensory awareness. Dreams, for example, do not create space-time. They are not explicated as matter/energy in our three-dimensional universe. It is quite possible, therefore, that sentient beings gain information in the second way, from beyond our normal senses. According to Seth, our identity does not reveal itself entirely in space-time coordinates. There are other means.

Seth (and for that matter Jung) sees the consciousness of a human being divided into inner and outer egos. The outer ego is aware of information coming through the senses and the nervous system; it is our normal consciousness. While most of the information given to the outer ego is sense data, other forms are available to us if we shift our focus. Intuitive knowledge comes in flashes of inspiration, insight, and spontaneous creativity, exhilarating moments that we all have experienced at one time or another. In such instances, the outer ego focuses outside the explicated order, and no energetic signal is required. This information is not given in explicated form and does not enter our three-dimensional world. It does not require energy for its transmission and is not subject to the explicated rules such as the special theory of relativity.

An analogous picture can be created for all consciousness, including particle consciousness. The information that binds all explicated particles together and makes them all part of one whole can be termed nonenergetic information. All points in space-time have immediate access to this vast storehouse, since nonenergetic information does not travel through space-time. When a particle "jumps" from one quantum state to another, there is no flow of energy. The energy used at each point is the zero-point energy normally available in space. The particle is explicated at one point and then at another; it does not pass in between.

To reiterate, we receive two types of information. The first is that gained through our senses (extended through instruments). The second is information that is not explicated or energetic but is available through a different focus of the mind, a focus that takes us into the implicate order and beyond. This latter type of information has the feature of nonseparability mentioned in the Bell-Aspect criteria.

In *Faster than Light*, Herbert expresses a different perspective. He points out that in 1978 physicist Philippe Eberhard developed a proof showing that all faster-than-light effects encountered in quantum calculations cancel out before the final results are obtained; all quantum predictions obey the tenets of special relativity. In contradistinction, Bell's theorem indicates that faster-than-light connections are real and are not just a theoretical artifact. To Herbert, the conflict is resolved: they are both right. Eberhard's and Bell's interpretations are applicable to different aspects of quantum measurement. Bell's theorem deals with individual quantum jumps, while Eberhard's proof is concerned with patterns of quantum jumps. Herbert uses the analogy of tossing dice. A single roll is completely unpredictable (random), but a large number of rolls develops a pattern. The pattern obeys the limit on light velocity. The individual toss is completely random and cannot pass information since random events cannot provide a medium for sending an intelligible message.

Herbert's views can be interpreted as follows. Information requires a pattern in our three-dimensional world. The pattern is composed of innumerable quantum jumps. The only way our senses can recognize information is by being exposed to a pattern formed by a large number of explications. This approach seems consistent with the relationship between the ideas of Eberhard and Bell. In other words, our senses follow Eberhard's patterns, and our intuition taps the implicate order.

According to the EPR experiment and Bell's inequality, the idea of an object-filled world is incompatible with the concept of nonlocality. If we have nonlocality, we cannot have a world of metaphorical billiard balls as described by the classical metaphor. Bohr chose locality and made reality observer-dependent. In developing a new metaphor for reality, what can be said about the existence of particle B if a measurement is not made? Assuming that particle B has a kind of consciousness (as we have been assuming here), then the consciousness of the particle explicates its own reality relatively independent of the observer. As far as the particle is concerned, it has an existence. However, it can explicate its own reality only within the rules of the game determined by all consciousness. That is, while particle B sees itself as a separate identity, it is not separate from particle A. Furthermore, it does not exist in the reality of the physicist who might measure it. Its reality conforms with the fact that if it were to be measured, its effects would be what the physicist expects from quantum theory. After all, the physicist never really sees the particle; it is inferred from a pointer on a meter (or a printout on computer paper). So if reality is made up of a malleable landscape waiting to be brought into focus, the particle focuses so that it seems to have real existence, and the physicist focuses so that the meters give the correct answers. Both realities mesh and are in conformity with each other, and the realities are nonlocal. The classical metaphor, then, is no longer valid.

9

Dissipative Systems: Nature's Creative Expressions

A system is said to be in equilibrium when it is balanced or stable, exhibiting an inert and supposedly completely random uniformity. A system that is near equilibrium is slightly out of balance; small changes in energy produce correspondingly small changes in the system. Generally, such small changes are reversible; that is, the initial state can be retrieved. (Classical physics largely concerns itself with systems near equilibrium, and the mathematics required is easily accessible.) Systems far from equilibrium are markedly out of balance. Even a small amount of additional energy can cause a large change. In fact, a given amount of added energy could cause a total reorganization of the system.

Systems can be defined as either "open" or "closed." An open system is one in which an exchange of matter and energy takes place with the outside environment. A closed system is an ideal system having little or no exchange with its environment. The entropy of a system is the amount of energy unavailable for work. As an example, a high-entropy system would be in thermodynamic equilibrium and would have a uniform temperature distribution, thus preventing heat energy from being utilized. The second law of thermodynamics states that in a closed system, entropy never decreases. In fact, the system *must* be closed for the second law to apply.

In 1977, after some twenty years of research, Ilya Prigogine received the Nobel Prize in chemistry for his extensive and detailed study of systems that are far from equilibrium. He called his radically new approach the theory of dissipative structures. The theory states that if a system is driven far from equilibrium by an insertion of energy from the outside, that system undergoes changes which, at a critical point, completely destroy the previous structure. In many cases, the system can then develop a new and different level of order. Prigogine called systems capable of this activity dissipative structures because they require an insertion of energy to sustain them. A closed dissipative system normally acts in a simple dynamic way: in the absence of outside influences, it comes to rest at one or more points of equilibrium. Conversely, if its initial condition is in equilibrium, it will remain so unless acted upon by outside energy.

Dissipative structures have two features of interest to our discussion. One is the concept of the bifurcation point. The other is that dissipative structures evade the second law of thermodynamics.

The bifurcation point is that critical point beyond which a new state can appear when energy is inserted into the system. As an example, let us say that before bifurcation a system is chaotic. At the point of bifurcation, predictability seems to collapse; that is, several choices emerge, but there appears to be no way to determine which course a system will take. A system seems to choose whichever probability it wishes to activate, and then the system reorganizes itself. (After bifurcation, a form of order might or might not be apparent.) Prigogine calls this attribute of far-from-equilibrium systems self-organization.

The bifurcation point was foreseen by the nineteenth-century physicist James Clerk Maxwell. Maxwell noticed unstable states existing in nature at points where deterministic laws were not applicable. He called these "singular points"; an example is a system in which a small cause creates a large, unpredictable effect. Note the following comment by Maxwell:

> If . . . those cultivators of physical science . . . are led in the pursuit
> of the arcana of science to study the singularities and instabilities,
> rather than the continuities and stabilities of things, the promotion

of natural knowledge may tend to remove that prejudice in favor of determinism which seems to arise from assuming that the physical science of the future is a mere magnified image of that of the past.[1]

Prigogine uses the term "active matter" in referring to the apparent capacity of matter for conscious or deliberate action. The material level, then, has its own will; it can mold itself into myriad forms and shapes. Since this process can occur in inanimate systems, the gap between the inanimate and animate becomes obscured. While classical physics viewed the world as a machine and therefore passive, Prigogine sees nature as active, as primarily self-organized. The similarity between Prigogine's views and Bohm's implicate and super-implicate orders is apparent. The essential feature of Prigogine's approach is that order can be created out of chaos.

In Bohm's terms, the chaos is not truly chaotic but is instead an order of higher degree. Recall that an order of high degree requires a large amount of information to describe it. If outside energy is inserted into a system with order of a high degree (nearly chaotic), and if the inserted energy is an order of low degree (very orderly), then the system can be converted from seeming chaos to orderliness. This is analogous to the superimplicate order (outside energy of a high degree of order) forming the complex implicate order into what we consider our very orderly three-dimensional world.

The second basic feature of dissipative structures that is of special interest here is that these systems evade (but do not violate) the second law of thermodynamics. Since the system is not closed, it escapes the second law by injecting entropy into its surroundings. The newly created system can be indefinitely sustained because it is constantly supplied with new injections of energy, or negative entropy.

The self-organization process can also spontaneously create order out of chaos. A system high in entropy (or disorder) can organize itself into complex forms and structures as long as it is not closed. Again, the similarity to Bohm's ideas is evident. Information comes into a disordered system and produces a structure with more order. The context is enlarged, and order is the result. The system is not bound to develop in predetermined paths wholly dependent on the initial conditions. A creative element is at work in nature that has the power to produce complex forms and structures. Physicist Louise B. Young sees in all this a universe creating itself out of the original disordered matter and radiation of the big bang. She says:

> I postulate that we are witnessing — and indeed participating in — a creative act that is taking place throughout time. As in all such endeavors, the finished product could not have been clearly foreseen in the beginning. We know from the experience of many artists, poets, and musicians that creativity is an uncertain reaching out, a growth, rather than a foreplanned or preconceived activity. It starts with something as vague as a need, an inspiration, a yearning. Poet Stephen Spender speaks of "a dim cloud of an idea which I feel must be condensed into a shower of words."[2]

As we noted earlier in connection with Bohm's thoughts regarding time, Prigogine sees time as irreversible. He connects that view to the idea of breaking the symmetry of a timeless order. At the bifurcation point, symmetry is broken in the transition to a more complex system, and time becomes irreversible. This is analogous to the collapse of the wave function in quantum theory. The timeless order of the wave function is changed to the irreversible timed order of our three-dimensional universe when a selection is made. The world we experience constructs time from its own process of creation.

The similarity between the universe and a dissipative system is clear. That is, structure and form were not present at the time of the big bang. There has been an evolutionary process, from the disordered system at the beginning to the highly structured system we now experience. A self-organizing or creative force in nature results in the highly ordered universe we now observe. However, according to the classical view, the universe is a closed system: the second law applies, the cosmos is progressing inevitably to a slow death, pockets of form and structure exist, but they too will eventually lose their source of energy. In contrast, Bohm, Wilber, and Seth all postulate that the universe is embedded in a series of orders and that our world is but a ripple in a vast ocean of energy. The universe is constantly receiving energy — from the holomovement (Bohm), deep structures (Wilber), or frameworks (Seth) — consistent with Prigogine's concept of the dissipative system. There is a constant rapid insertion of energy into our universe and a constant rapid expulsion of energy out of our world. From this perspective, the three-dimensional world can be viewed as a dissipative structure with a self-organizing principle. The self-organizing principle can be identified with the superquantum potential of the universe. We noted earlier that a closed dissipative system at equilibrium will not change from its position of rest without outside influences. The self-organizing principle, then, must originate outside the system. Prigogine finds that there seems to be communication between molecules over long distances and long periods of time. That is, molecules, which are supposed to move in accordance with causal laws and respond only to local forces from neighboring molecules, seem to act collectively and in a purposeful manner, far beyond the area of local forces. This is true even for so-called nonliving systems.

Prigogine has brought the uncertainty of the quantum world into the macro world. Before Prigogine's work, it could be said that while probability played a major role on the particle level, the large number of particles in a macro system completely eliminates uncertainty on the classical plane. With the bifurcation point in large-scale far-from-equilibrium systems, probability enters the everyday world. Nonliving systems are seen to behave in a wholistic manner similar to biological systems. Our concept of the barrier between life and nonlife appears to be disintegrating, due in part to Prigogine's theories. Perhaps the final breakdown will occur when the so-called non-life material is seen to be alive, having its own level of awareness.

10

Eastern and Western Thought as Aspects of a Common Reality

*I*n classical terms, a particle is defined as "stuff" confined to a small space, while matter waves are seen as fields spread throughout space. According to quantum theory, a complete picture of an elementary particle requires that it be viewed sometimes as a particle and sometimes as a wave. Both interpretations are suitable, depending on the situation. Both views are necessary to describe the reality of a particle, and both are, as Bohr said, complementary to each other.

COMPLEMENTARITY

The link between particle and wave becomes extraordinary when their relationship is defined more exactly. Quantum theory and experimental data tell the physicist that when a quantum particle is measured (observed), it acts like a particle. When the quantum particle is not being measured, it acts like a wave. The wave is "seen" only in the pattern that an aggregate of particles produces under certain defined conditions. If the particle were assumed to be only a wave, physicists would have to discard the evidence of all the observations that show it to be a particle. On the other hand, if the quantum particle were assumed to be a particle in the classical sense, then electron diffraction[1] could not be explained. We are left with "quantumstuff" existing as a particle in our three-dimensional world and appearing only when we observe it, and otherwise, existing as a wavelike entity not in our three-dimensional world. Hence, Bohr's solution: in describing basic quantum reality, a complementary relationship exists between a particle and a wave. Quantumstuff inhabits our world and also exists in some other world in a different guise.

David Bohm found a model for this strange situation when he applied his causal interpretation to quantum field theory. He postulates that the particle and the wave are actually aspects of the same thing or, more accurately, the particle is an explication, or wavelike ripple, from the implicate order. Furthermore, Bohm brings in the observer through the concept of the superimplicate order. The implicate order is basically passive, or linear, in Bohm's terms. It does not have the intrinsic capacity to unfold into the explicate. The superimplicate order produces nonlinearity in the implicate order and thus unfolds it into forms and structures. The superimplicate order can be seen as analogous to Seth's Framework 3, where sentient minds reside, and as the home of the "mind" (superquantum potential) associated with the particle.

Seth introduces the wave-particle concept when he discusses the basic consciousness units (CUs). He comments:

> CU's can also operate as "particles" or as "waves". . . . When CU's operate as particles, in your terms, they build up a continuity in time. They take on the characteristics of particularity. They identify themselves by the establishment of specific boundaries.
>
> They take certain forms, then, when they operate as particles, and experience their reality from "the center of" those forms. They concentrate upon, or focus upon, their unique specifications. They become *in your terms* individual.
>
> When CU's operate as waves, however, they do not set up any boundaries about their own self-awareness — and when operating as waves CU's can indeed be in more than one place at one time.[2]

The basic constituents of reality seem to be capable of expression as both particle and wave. When operating as particles, they become individualized and display the attributes described as matter. They are projected and injected to and from our three-dimensional universe and take on a timed order. At the very basis of reality, then, the concept of wave-particle complementarity is essential. This may explain why complementarity finds expression in the mathematical formalism of quantum theory without creating contradictions.

WHOLISM

Since Galileo, experiment has served as the heart of physics. In order to test the promulgated laws of nature, physicists performed experiments to enable them to describe the structure of matter. That is, they analyzed, dissected, and separated matter into ever-smaller fragments, subjecting it to myriad forces so that they could extract its fundamental attributes that remain the same regardless of change. The goal of all these experiments was to understand the basic constituents of matter. The belief was that when this fundamental element was understood, a clear picture of the whole could be constructed through the sum of its parts. But when the fragmentation became extremely small (quantum particles), nature began responding in unexpected ways. Separating or isolating an object from its surroundings can produce surprising, counterintuitive results. As the naturalist and writer John Muir said, "When we try to pick out anything by itself, we find it tied to everything else in the universe." Breaking the world into smaller and smaller parts exacts a price by limiting our knowledge. Nevertheless, the elemental approach of classical physics has been quite successful in producing an advanced technological culture. It is doubtful that a more wholistic approach, without the separations created by modern experiments, would have achieved the same results.

Nature itself seems to have faced a similar situation in designing the brain. The two hemispheres of the brain appear to duplicate each other when casually examined. Yet research shows that they are functionally distinct. The left side is normally referred to as the logical, digital side, while the right is seen as using an analogue approach. The left is sequential, and the right is wholistic. It is almost as if we had in our heads both Democritus, the reductionist, and the wholistic Aristotle. Democritus taught us to reduce information to its basic elements and through analytical and logical sequences build up the whole. Aristotle believed in seeing the total picture at once. We have one consciousness but two ways of processing sensory information, which gives us a more complete picture. Thus, the two sides of the brain reflect reality in complementary ways.

Remarkably, each hemisphere is neither completely independent from the other nor are they totally fused. Recent research shows that various mental functions are dispersed throughout the brain, often in a redundant manner. Still, the primary mode of functioning seems to be that each side reacts in its characteristic way in a mutual arrangement with the other side, and the arrangement changes quickly as the need arises. In this manner, the brain is able to process information from differing aspects and angles. The right side provides context and order, while the left provides facts and detail. Both are needed for a balanced life, and the whole is far greater than the sum of its parts.

MIND AS A TOOL

The collective human consciousness can be seen as similar to the brain in that it also seems to have a "left" side and a "right" side. The scientific view prevalent in the West is analogous to the left side of the brain, while the wholistic approach typical of Eastern thought is analogous to the right side. Unfortunately, the analogy is not exact. The two sides of the brain are connected by the fibers of the corpus callosum, whereas East and West have not yet found a bridge between their differing philosophical orientations. It is becoming increasingly clear, though, that a future scientific paradigm will need to incorporate both: reality cannot be satisfactorily explained by exclusively embracing either the Western or Eastern view.

We in the West may have reached an impasse using our particle accelerators. Perhaps we must now turn to other less expensive and less technologically sophisticated instruments. The East has a suggestion for us: creative use of the human mind. Mystics use the mind to probe a transcendental reality not accessible to the instruments of science, a realm that is the abode of the wave aspect of reality. After all, the mind is the instrument for introspective experiments, just as the particle accelerator is an instrument for three-dimensional experiments. Altered states of consciousness cannot be probed and examined with the methods and tools of Western science.

Can we envision a time when our physics departments will offer Meditation 101 along with Quantum Theory 101? Both may be needed to create an effective world model for the future. As Schrödinger stated:

> I do believe that this is precisely the point where our present way of thinking does need to be amended, perhaps by a bit of blood-transfusion from Eastern thought. That will not be easy, we must beware of blunders — blood-transfusion always needs great precaution to prevent clotting. We do not wish to lose the logical precision that our scientific thought has reached, and that is unparalleled anywhere at any epoch.[3]

And Fritjof Capra quotes physicist John Wheeler:

> One has the feeling that the thinkers of the East knew it all, and if we could only translate their answers into our language we would have the answers to all our questions.[4]

PHYSICAL SCIENCE AND INNER SCIENCE

In Seth's view, each probability system (such as our three-dimensional universe) has what he calls inner working plans or blueprints. These blueprints present themselves in various ways: physically, psychically, spiritually, mentally. As an example, each individual has his/her own blueprint. Biologically, the information is "knit into the genes and chromosomes," Seth says, but it really exists apart from its biological manifestation, and the physical structure acts only as a carrier of the information.

> In the same fashion the species *en masse* holds within its vast inner mind such working plans or blueprints. They exist apart from the physical world and in an inner one, and from this you draw those theories, ideas, civilizations, and technologies which you then physically translate.[5]

Similarly, Plato saw a constant change in the material things surrounding him, but he saw the idea or the form of the thing as changeless, existing in a realm outside our three-dimensional world. Seth identifies this domain as Framework 2, or the implicate order.

Bohm disagrees with the Platonic view that these forms are perfect and unchanging. He sees them as engaged in a developmental process. Seth agrees with Bohm:

> Platonic thought saw this inner world as perfect. As you think of it, however, perfection always suggests something done and finished, or beyond surpassing, and this of course denies the inherent characteristics of creativity, which do indeed always seek to *surpass* themselves. The Platonic, idealized inner world would ultimately result in a dead one, for in it the models for all exteriorizations were seen as *already* completed — finished and perfect.

> Many have seen that inner world as the source for the physical one, but imagined that man's purpose was merely to construct physically these perfect images to the best of his abilities. In that picture man himself *did not help create* that inner world, or have any hand in its beauty. He could at best try to duplicate it physically — never able, however, to match its perfection in those terms. In such a version of inner-outer reality the back-and-forth mobility, the give-and-take between inner and outer, is ignored. Man, being a part of that inner world by reason of the nature of his own psyche, automatically has a hand in the creation of those blueprints which *at another level* he uses as guides.[6]

Since these blueprints for reality are not normally seen, and are indeed invisible, we deny or "forget" their existence. In order to "see" these idealizations, we must as a first step accept the fact that beneath our world lies another one, that our three-dimensional world is embedded in the inner world. We must also accept the notion that along with our normal focus of consciousness, there are other focuses that are also legitimate. William Irwin Thompson reminds us of the inescapable limitation of our perception:

> [Marshall] McLuhan was fond of saying, "I don't know who discovered water, but it certainly wasn't a fish." If we are inside an all-encompassing environment, we do not see it.[7]

Seth uses the analogy of a photograph:

> In your terms a photograph freezes motion, frames the moment . . . that you can physically perceive.

In usual circumstances you may remember the emotions that you felt at the time a picture of yourself was taken, and to some extent those emotions may show themselves in gestures or facial expression. But the greater subjective reality of that moment does not appear physically in such a photograph. It completely escapes insofar as its physical appearance within that structure is concerned. In the same way the past or the future is closed out. The particular focus necessary to produce such a picture then necessitates the exclusion of other data. That certainly is obvious. Because you must manipulate within specific time periods, you do the same kind of thing in daily life, and on a conscious level ignore or exclude much information that is otherwise available.[8]

The information based on the inner blueprints is normally not available to our science as currently constructed. The instruments we use for research are useful only in measuring the level of reality in which they themselves exist. In a sense they interpret the universe in horizontal terms. In studying the deeper universe they are all but useless, and may even be misleading. Seth does not say that the use of instruments is futile, but that it is limiting. After all, we in the West have advanced in a technology that we particularly desired; we traveled to the moon, we have conquered many diseases. But in Seth's terms, we must realize some day that we are facing a stone wall. People will continue to die of other "unconquered" diseases, for example; a person ready to die will do so, despite any medication. How are we to understand this? Seth explains:

> The particular thrust and direction of your own science have been directly *opposed* to the development of such inner sciences . . . [which means that] to some extent each step in the one direction has thus far taken you further from the other. Yet all sciences are based on the desire for knowledge, and so there are intersections that occur even in the most diverse of paths; and you are at such an intersection.
>
> Your own science has led you to its logical conclusion. It is not enough, and some suspect that its methods and attitudes have a built-in disadvantage. Physicists are going beyond themselves, so to speak, where even their own instruments cannot follow and where all rules do not apply. Even the prophet Einstein did not lead them far enough.

You cannot stand apart from a reality and do any more than present diagrams of it. You will not understand its living heart or its nature.

The behavior of electrons, for example, will elude your technological knowledge — for in deepest terms what you will "perceive" will be a facade, an appearance or illusion. So far, within the rules of the game, you have been able to make your "facts" about electrons work. To follow their multidimensional activity however is another matter — a pun — and you need, if you will forgive me, a speedier means.

The blueprints for reality lie even beneath the electrons' activity. As long as you think in terms of particles, you are basically off the track — or even when you think in terms of waves. The idea of interrelated *fields* comes closer, of course, yet even here you are simply changing one kind of term for one like it, only slightly different. In all of these cases you are ignoring the reality of consciousness, and its gestalt formations and manifestations. Until you perceive the innate consciousness behind any "visible" or "invisible" manifestations, then, you put a definite barrier to your own knowledge.[9]

Seth suggests that we start training mental physicists and work at developing a nonsensory physics. These physicists would study and map the implicate order, but not with the use of microscopes and accelerators; they would have a different type of training and use different tools. This new physicist might study dreams, as an example, but *not* by studying the dreams of others in a laboratory, and *not* by measuring the physical changes that occur during the dream state. The mental physicist might first learn to become conscious in normal terms while in the sleep state, as in lucid dreaming. He or she could then learn to recognize many different kinds of reality, isolate them, and attempt to identify the laws that govern them. This new scientist might get a glimpse of the blueprints for our three-dimensional world. With our present science *and* an inner science, we could understand the workings of both our exterior universe and the interior one from which it springs. As Seth points out (with reasoning similar to Gödel's), perhaps we must go to another level of consciousness to understand this one. We could then see the three-dimensional world as if it were a photograph in our hand. We could see it from a broader context that would more truly and more fully reveal its nature.

In his book *Star Wave,* Fred Alan Wolf recounts an experience that illustrates this point. While a visiting professor at Birkbeck College at the University of London, he had a vivid lucid dream, which he describes as follows:

> As I descended I became more and more aware that I was dreaming. It was dawning on me that I was both snuggled cosily in bed and slipping through space-time in a dream of uncanny proportions. It was as if my awareness were split in two. To my great surprise I was conscious that I was asleep. What a contradiction! How can you be asleep and conscious at the same time?

> Next I found myself awakening, but I was shocked to discover that I had not actually awakened at all: I was dreaming that I was awakening and I knew it! No sooner had I realized that I was still dreaming than I would awaken once more from the dream to dream that I was awakening once again. It was like ascending through a set of Chinese boxes: As soon as I was out of one box I found that I was inside of another, still larger box. I soon realized that I was in control of my dream. I could awaken for real or I could descend to any universe layer I wished and experience my dream consciously. . . .

> It is very important to realize that this "dream" was not just an ordinary dream. I was fully conscious not only during it but in the transition from the astral plane to my bed.[10]

Wolf later went with friends to a Druid ceremony. His comments are particularly interesting:

> [The ceremony was] presided over by the then chief Druid, Dr. Thomas Maughan. . . . After the service Nancy asked me to speak up and tell Maughan about my dream, and I did. Maughan listened attentively. When I finished, he looked at the group of attendees and reaffirmed that this was no ordinary dream. He asked me what I did for a living. I told him that I was a visiting professor of physics at Birkbeck College. Then, astonishingly, he admonished me for not being more attentive to details when I was there. "You weren't a too careful physicist. . . ."

> And he was right. I was so struck that I had actually gone to the astral level and had had the experience that I failed to take it into account as an experience worthy of physical laws and subject to the

same scrutiny as any other physical experience. My own scepticism had defeated my acceptance of its reality. To my rational mind "it was only a dream."[11]

Had Wolf been a physicist of inner science as well as our usual science, he may have discovered some aspects of the inner blueprints.

AN UNDERLYING UNITY

We have been discussing aspects of physics, mysticism, and the occult, with emphasis on the similarities among them. If there is one central common concept to all these disciplines, it is that of wholeness. The universe cannot be viewed as a group of separate entities operating on each other through physical contact but must be seen as a complete organism, moving and changing through an organized, systematic series of events directed toward a common purpose. With this vision, the individual entity retreats into the background, and the underlying relationships become primary. In a similar manner, physics, mysticism, and the occult are not separate entities brought together by some causal arrangement; rather, each one is contained within the others.

Perhaps the mystical and the occult are embedded in modern physics as random background noise that bears within its ostensible chaos the strange ideas that Niels Bohr found so exciting and so necessary. Such revolutionary ideas must be allowed to emerge and be examined, either to be incorporated into our new scientific outlook or to be placed back into the background to await a more propitious moment. The philosophical rigidity of contemporary physics needs to be energized by the "unthinkable" ideas common in Eastern thought. Science as epitome of Western thought and spirituality as the epitome of Eastern thought are not separate disciplines; they are different aspects of the same whole, each dancing around the other, waiting to be merged in human awareness.

11

Toward A New Paradigm

*I*n 1980, Stephen Hawking assumed the prestigious position of Lucasian Professor of mathematics at Cambridge. His inaugural address, entitled "Is the End in Sight for Theoretical Physics?", proposed that "we might have a complete, consistent, and unified theory of the physical interactions which would describe all possible observations." Hawking considered this goal achievable within the lifetime of some persons now living.[1] If Hawking's optimistic prediction is correct, would this be a "theory of everything," as some physicists suggest?

Clearly such a theory is not consonant with the views expressed here. It is the position of Bohm, Wilber, and Seth that the discipline of physics as presently constituted is an abstraction of a limited number of elements from an infinitely complex, fluctuating, and therefore indefinable background. That would mean that any theory of everything can only be a doorway to a new level of greater subtlety and a more encompassing realm. Hawking and Bohm obviously exhibit different metaphysical outlooks.

As Thomas Kuhn notes in *The Structure of Scientific Revolutions*, most scientists operate within a fixed conceptual framework. If this framework attracts the major portion of the scientific community from competing modes of thought, and if it is open-ended enough to present problems to solve, then it functions as a paradigm. Such a paradigm becomes a lens through which experience is seen, interpreted, and correlated. The agreed-upon overview provided in this way by a paradigm allows scientists to engage in what Kuhn calls "normal" science, without the handicap of having to reexamine basic assumptions in relation to each new idea.

Scientists working within a given paradigm are all committed to the same rules and accept the same norms in performing their research. In this way a paradigm becomes a filter through which experiences are passed to determine their acceptability within the framework of the scientific discipline. If an experience cannot pass through the filter, it usually is discarded and considered nonfactual, not real. An explanation of the experience is then not required. As Abraham Maslow has observed in this connection, "To him who has only a hammer, the whole world looks like a nail." Thus, any theory, any model, any discipline of study is built on a limited number of elements extracted from the totality of information. When anomalies arise, one does well to remember that only a part of the whole has been considered.

Although Kuhn's work deals strictly with science, the concept of paradigms — especially paradigm shifts — resonates in other areas. The scientific paradigm, at least in a general way, is a major component of the metaphysical outlook of society as a whole. At the societal level, for example, we see signs around the world that the old, accepted, familiar ideas that have informed human lives no longer seem to fit — just as Newtonian laws no longer describe subatomic reality. Many observers of Western society have voiced their concerns that not only is our present scientific paradigm too narrow, but that it is time for a broadening — and deepening — of the general worldview. We would expect that a change in the fundamental framework employed in scientific investigation would make itself felt throughout human culture, in precisely those areas where we now sense that our old philosophies no longer serve.

THE EVOLUTION OF PARADIGMS

Accepting a new paradigm is like acquiring a new wardrobe. Initially, the garments fit well, look stylish, and are suitable for almost all occasions. However, with the passage of time, the clothes become too loose or too tight, frayed and tattered, and the wearer begins to feel unsuitably dressed for certain events. At this point, he or she can either alter the outfits or purchase a new wardrobe. But the older clothes are not so easily cast aside. They are more comfortable in some ways, they served long and well, they are like old friends; a certain attachment has set in. Indeed, the wearer may decide to keep the old wardrobe and restrict his or her activities accordingly, passing up occasions at which the clothes seem out of place. The activities that are dropped become defined as unimportant and eventually no longer belong to the wearer's "real" world. Choosing a new wardrobe, on the other hand, is comparable to what Kuhn terms a paradigm revolution: the basic framework that defines activity is altered, and "normal" science is replaced with a new range of possibilities. We may call this a new "reality."

Up until the time of Newton, the European worldview was defined metaphorically by the garments of the church. In time, this wardrobe grew outmoded. The reality of the supernatural was discarded along with the stifling authoritarianism of the church and other aspects of ecclesiastical garb. So was the reassuring promise of immortality.

The Newtonian Model

One of the basic notions of nineteenth-century physics is that time, space, and matter are all independent of each other. Matter was seen to inhabit space, but not affect it, just as we move about in our house but do not change its structure in so doing. In a similar sense, time was thought to flow forward at a steady rate, completely unaffected by matter and its activities in space. With a steady flow of time in the background, concepts such as velocity and acceleration were firmly established. Space was seen as a receptacle inside which matter had a definite position and form. All was predictable and orderly. The nineteenth-century physicist could deal with space, time, and matter in a coherent and consistent way.

Science, we might say, had set the fashion with the appealing new styles of reductionism and determinism. But the individual felt alone, without recourse to traditional religious beliefs, a mere cog in the proverbial machine. Those who clung to the old garments of faith were labeled with the heresy of being "unscientific."

Relativity and Quantum Theories

The Newtonian worldview, when assessed from the perspective of physics, still has a great deal of relevance. Relativity theory comes into play only when particles have a velocity near the speed of light, and quantum theory becomes necessary largely in the micro world. On the macro scale, Newtonian mechanics is still the algorithm of choice. NASA's rocket program, for example, relies much more on Newton than on Einstein. However, the philosophical foundation of Newtonian physics was shaken in a fundamental way around the beginning of the twentieth century.

Since a central tenet of relativity theory is that a signal cannot travel faster than light, the notion of a rigid body had to be discarded. The concept of small, hard balls moving through the void of space was no longer tenable. Relativity theory also meant that the absolute and even flow of time was rejected in favor of a time flow dependent on the speed of a coordinate frame. In addition, the smooth continuum of the Cartesian world was called into question by quantum theory; matter and energy seemed to display a grainy structure. Under the influence of Einstein's theories, then, the empty, linear container of space began to bend and change. And under the influence of quantum theory, space appeared choppy and uneven — in John Wheeler's words, a churning foam.

Furthermore, quantum theory tells us that space is not a vacuum but a plenum, an ocean of energy whose density far exceeds that of the elementary particle. In short, matter appears to be small perturbations causing distortions in the ocean of space, and space is no longer considered a smooth, flat, lifeless container; it has structure. The foundation of the universe, then, seems to be not at the level of particle interaction but at an even more basic level. Both Wheeler and David Bohm propose that the guiding principles of our material universe must be sought in the plenum.

The Copenhagen Interpretation

The predictive ability of quantum theory is uncanny. It has enabled us to create computers and lasers. And yet it starts from randomness; it does not make intuitive sense. For this reason, the

Copenhagen interpretation — which stands as the current paradigm for contemporary physics — fails to offer a useful metaphor for reality.

Bohr changed Heisenberg's uncertainty to ambiguity, explained by Bohm as follows:

> Consider what is meant by the term *temperature*. Temperature, as measured by a thermometer suspended in the air, is in fact a measure of the mean energy of the air molecules. It is essentially a statistical concept which has a clear definition when a very large number of molecules are involved. But what is the meaning of the temperature of a single molecule, or for that matter of a single atom? Clearly the concept is by no means *uncertain*; rather it is *inherently ambiguous*.[2]

Bohm goes on to describe the Copenhagen interpretation:

> The *entire phenomenon* in which the measurement (or any other quantum measurement, for that matter) takes place cannot be further analyzed into, for example, the observed particle A, the incident electron, the microscope, and the place at which the spot Q appears. Rather the *form* of the experimental conditions and the *content* of the experimental results are a *whole* which is not further analyzable in any way at all. In the case of the microscope, this limit to analysis can be clearly seen, for the *meaning* of the results depends upon the way in which the spot Q and the particle A are linked together. But according to the laws of quantum theory, this involves a single quantum process which is not only indivisible but also unpredictable and uncontrollable.[3]

In summary, as Bohm points out, if Bohr were asked if an electron existed, he would answer that the question has no meaning except as "an aspect of the unanalyzable pattern of phenomena in which its observation takes place" [Bohm's paraphrase]. That essentially means that the measuring equipment and the electron cannot ever really be understood. Quantum theory, then, is a procedure for solving problems: it predicts how measuring devices will reflect the activities of the quantum world. But it has nothing to say about the characteristics of the quantum world. It is not a model because it does not explain anything. Perhaps, though, we can hope to uncover a vision of nature more encompassing than the old mechanistic order, in which quantum theory and relativity theory will be seen as more than mere algorithms.

The new wine of quantum theory, then, is contained in the old wineskin of the concepts and language of classical world outlook, resulting in the paradoxes that plague quantum physics. Nevertheless, Bohm has made a start in establishing a new foundation for physics. Incorporating the contributions of Wilber and Seth, we can try to expand his concepts to a vision that encompasses all disciplines.

ELEMENTS OF A NEW VISION

In the introduction, we discussed the possibility of discerning common elements among the views of Bohm, Wilber, and Seth that could form a new metaphor for reality. Such a metaphor, like any other, must necessarily be incomplete and open to revision. Still, these elements may constitute the beginnings of a new paradigm for science and a new way of understanding reality.

Consciousness

One of the most significant contributions to contemporary physics was Einstein's discovery of the equivalence of mass and energy. A fundamental equivalence we have seen in this examination is that of matter/energy and consciousness. This means that all existence is a form of consciousness and is alive in some sense; inert or dead matter is an illusion. As Seth points out, "Scientists say now that energy and matter are one. They must take the next full step to realize that *consciousness* and energy and matter are one."[4]

In today's normal science, the central problem involving consciousness is expressed in this form: At what point in the structure of inanimate matter does consciousness appear? (The answer to this question is crucial to the field of artificial intelligence.) In the new paradigm, the question is, how does matter originate from consciousness? Consciousness is understood to be manifest in a particular gestalt at each level, with matter created through the willing cooperation of all forms of consciousness, from the human to the nearest rock.

Levels of Consciousness

The configuration of matter is determined by the level of consciousness. Since the possibilities for configuration are infinite, consciousness can never be fully defined.

Bohm incorporates these views in his model of the implicate and the superimplicate orders. In his concept, any mass is a manifestation of a matter field. A particle is related to a quantum potential, which exhibits its activity not through its intensity but through its information content. The quantum field and its manifestation, the particle, willingly cooperate to perform the dance we call physical reality. The field represents the implicate order, which is enfolded. The particle represents the portion that is unfolded, the explicate order. The unfolded portion is our material universe; it springs from enfolded order and gives us the experience we call perception.

The implicate order has another significant characteristic: it is passive. An organizing element is needed to unfold it, what Nick Herbert refers to as the choreographer. Therefore, a third level is required that performs this function. In Bohm's terms this is the superimplicate order, in Seth's terms it is Framework 3. In Wilber's terms it is the self on the next higher level.

The first level is the explicate order: our everyday world, where physics normally plies its trade. The anomalies and paradoxes that we see in nature are the result of our restriction of scientific investigation to the explicate order. That is, although our universe performs its function in the arenas of time and space, *time and space are not primary*: there exists an underlying realm. Whether we see space as an infinite ocean of minute fluctuating black holes or as a multidimensional vibrating string is not important; the fact is, the classical view of space has become insufficient if we want to look more deeply into reality. The openings through to the implicate order are there — at the incredibly short distances of the Planck length.

The second level, the implicate order, contains all possibilities and probabilities. In this region, consciousness takes the form of waves rather than particles. The implicate order is whole, seamless, unbroken. To use a musical analogy, the implicate order contains all the possible music to be played.

Level three is where, to pursue the musical analogy, the conductor performs and where the selections from level two are made. The music played on level one is the result of an interplay of all the levels. If this music sometimes seems chaotic, it may be, as Bohm says, that the randomicity is an illusion caused by the limited context of our scientific views.

Mind-Body

As we saw in Chapter 1, the concept of a three-level world can be enlarged to include an infinite spectrum of consciousness. In Bohm's view, the relationship of the superimplicate order to the implicate order is similar to that of consciousness to matter. In the superimplicate-implicate relationship, the superimplicate is the conscious aspect, while the implicate is the material aspect. This relationship led Bohm to the principle of soma-significance, which states that any process can be viewed from its soma (material) aspect or its significance (mind) aspect, but the fact remains that there is really only one system. In the case of human beings, the brain and the mind are seen as related in a soma-significance manner.

The same idea is reflected in Wilber's deep structure and surface structure. The deep structure can be viewed as significance and the surface structure as soma. However, remember that the deep structure cannot manifest itself as a surface structure without the choreographer of the self. To accomplish this manifestation, the next-higher-level deep structure takes on a significance role and organizes the lower deep structure, which then takes on a soma role. The result is the explication of a surface structure, presumably for the benefit of the self, which is residing on the next higher level.

Motion

The classical approach to motion is embodied in Cartesian coordinates and implies a localized entity traveling through space-time. Bohm rejects this description and instead makes a distinction between what is manifest and unmanifest, leading to the notion of enfolding and unfolding as basic to motion. The unfolding is a projection of the implicate order to the explicate, and the enfolding is an introjection of the explicate back into the implicate. This idea finds common ground with Wilber's concept of microgeny and Seth's descriptions of pulsations. The unfolding and enfolding process requires an underlying order and cannot be explained by the Cartesian/Newtonian view since they assume only one level of reality.

When matter "moves," it does so as a series of projections and injections that create the illusion of motion, just as lights along a string quickly flashing on and off in succession look like a single light moving through space. In the particle world, this concept is more evident. In measuring an electron, when it moves from A to B, a measurement is made at A and then at B. The electron's appearance at consecutive points suggests that movement took place. But in Bohm's view, there was a projection and injection at each point.

Matter Formation

Another aspect of the unfolding and enfolding process is the creation of matter (discussed in Chapter 5 using the concept of black/white holes). Seth says that the form taken by matter arises from a series of pulsating projections and injections that has come into a sort of balance, or closure, and is stable. The frequency of these pulsations is much too rapid for our senses to follow; as a result, matter seems to have solidity and permanence. According to Seth, the forms are created from EE units (Chapter 3) emanating from various levels of consciousness. The EE units, in turn, use coordinate points as transformers to produce the forms that are symbols for our thoughts and feelings. This idea goes beyond Bohm's implicate order concept, but is not inconsistent with it.

Multidimensional Space

One of the most mind-boggling concepts promulgated by Seth is multidimensional space. Essentially Seth says that all physical objects are symbols created by a given level of consciousness, that each object or symbol is a product of a gestalt of consciousness, and that each consciousness requires its own three-dimensional space. In other words, we each create and live in our own universe, but our realities are correlated in such a fashion that intercourse between consciousnesses exists so that learning can take place. We have suggested that Seth's interpretation offers an explanation for the paradox of Schrödinger's cat (Chapter 7). Hilbert space, which is used as an algorithm in quantum theory, might be the counterpart of multidimensional space in contemporary physics.

Time

By definition, the implicate order is wholistic and undivided. Therefore, any connections in this order are independent of locality in space and time. Our three sources share the fundamental notion that events do not happen, they just *are*. Our sense of time and change comes from the unfolding of these events; history is the unfolding of an enfolded order; the past has been unfolded, the future is yet to be unfolded. But in the implicate order, all events are in the present. Seth says that one of the root assumptions of our current paradigm is space-time. Bohm says that our timed explicate order arises from a ground that does not exhibit time; the past and future are overtones of the continuous present of the implicate order.

Theories such as the big bang and evolution can be seen, from this perspective, as products of the explicate order only. They are explanations represented by the symbols of the unfolded order we call our universe. Some might suggest that if time is an aspect of a timeless order, then determinism is a valid notion. But that is not the case. Our world is made up of happenings that have been selected, and are yet to be selected, from a vast sea of probable events. We are not totally restricted in our choice; free will does operate. Since the present is all there is, our past and our future are constantly being created through the free choices of the individual consciousness involved. The decomposed wave function in quantum theory represents this freedom. More generally, many of the concepts and mathematical formalisms of contemporary physics undoubtedly have metaphoric relevance to an overall reality, but referring to them by the term "algorithm" reflects the scientist's reluctance to face this fact.

Purpose in the Universe

Alex Comfort notes that the objective world of elementary particles is simply a display of events, a false image giving the appearance of separate and distinct entities. Even if we accept that our space-time perspective merely gives that impression and no event is separate from another — all being interconnected through the multidimensional space, which is their source — still, a display implies a viewer. Why do we have a display, and who is doing the viewing? Physics, as presently constituted, must remain noncommittal. Seth, however, observes:

> In your system of reality you are learning what mental energy is, and how to use it. You do this by constantly transforming your thoughts and emotions into physical form. You are supposed to get a clear picture of your inner development by perceiving the exterior environment. What seems to be a perception, an objective concrete event independent from you, is instead the materialization of your own inner emotions, energy, and mental environment.[5]

Ken Wilber defines the purpose of the Atman-project as the self engaged in ascending the ladder of consciousness in order to produce ever higher unities. The goal is to bring all levels of consciousness to the top level. Ironically, the climb is an illusion. The higher self exists on all levels all the time; all boundaries are self-imposed. We must learn to widen our focus and thus rid ourselves of these boundaries.

Bohm also postulates a deeper purpose. Renée Weber recalls his comments from an interview:

> In one of our sessions, he suggests that what the cosmos is doing as we dialogue is to change its idea of itself. Our doubts and our questions, our small truths and large ones are all forms of its drive toward clarity and truth. Through us, the universe questions itself and tries out various answers on itself in an effort — parallel to our own — to decipher its own being.
>
> This, as I reflect on it, is awesome. It assigns a role to man that was once reserved for the gods.[6]

Light

As Einstein found, matter and energy are related by the velocity of light. For both the mystic and the physicist, light has an unmatched significance. *The Tibetan Book of the Dead* states that light is the only reality; all else is illusion. To Bohm, light (the entire electromagnetic spectrum) is more than energy as defined in physics; it is the potential of literally everything. Light enfolds the whole material universe; it is the implicate order. Light unfolds to form the explicate order. In this view, light can never be fully explained in terms of time and space and the explicate order only. As Seth continually affirms, our spectrum is but one portion of the total spectrum, and the totality is infinite. Bohm labels this totality the holomovement. So what we have as the basis of the new paradigm is an ocean of light energy, or the holomovement, which manifests itself as particles to create the explicate order.

Wholeness

Wholeness is vital to any new paradigm that emerges from these ideas; reality is not composed of separate elements of any description; everything is interconnected and interpenetrated. We can only extract certain aspects for examination in an approximate and restricted fashion. All That Is is whole and undivided.

Equally essential to our new paradigm is this: reality at its most fundamental level is consciousness. As Seth says, "All energy contains consciousness." If we accept that simple statement, it would indeed change our world. What we call matter is a gestalt of consciousness, and consciousness is light. So: All is light, and light is consciousness, and that is All That Is.

Notes

PREFACE

1. p. 17. Nick Herbert, "Scientists Explore Invisible Ocean of Glue," C-Life Institute, Boulder Creek, California, 1977, pp. 5-6.

2. p. 17. Quoted in Bob Toben and Fred Alan Wolf, *Space-Time and Beyond: Toward an Explanation of the Unexplainable*, new ed. E. P. Dutton, New York, 1982, p. 126.

3. p. 17. Arthur Hastings, an authority on channeling, said this regarding Jane Roberts and Seth:

   > Jane explored many possible explanations for the nature of Seth. She did not believe he was a secondary personality or part of the subconscious, nor did she want to refer to him as a spirit. She speculated that he might be a personification of the superconscious part of her self, a kind of psychological structure that enabled her to tune into revelational knowledge. She also allowed that he might have an independent existence as another entity. . . . Her honesty in facing this puzzle indicates both integrity and intelligence.

   From Arthur Hastings, *With the Tongues of Men and Angels: A Study of Channeling*. Holt, Rinehart and Winston, Inc., Fort Worth, 1991, p. 73.

INTRODUCTION

1. p. 19. Mendel Sachs, *Einstein vs. Bohr: The Continuing Controversies in Physics*. Open Court Publishing Company, LaSalle, Illinois, 1988, p. 264.

2. p. 21. John Archibald Wheeler, "Law Without Law," in John Archibald Wheeler and Wojciech Hubert Zurek, eds., *Quantum Theory and Measurement*. Princeton University Press, Princeton, New Jersey, 1983, pp. 192, 194.

3. p. 22. Banesh Hoffmann, *The Strange Story of the Quantum*. Pelican Books, Harmondsworth, England, 1959, p. 23.

4. p. 22. Quoted in Morris Kline, *Mathematics and the Search for Knowledge*. Oxford University Press, New York, 1985, p. 238.

5. p.23. Fred Alan Wolf, *The Body Quantum: The New Physics of Body, Mind, and Health*. Macmillan Publishing Company, New York, 1986, p. 257.

6. p. 23. Harold J. Morowitz, "Rediscovering the Mind." *Psychology Today*, August 1980, p. 12.

7. p. 24. Quoted in Harold J. Morowitz, *Cosmic Joy and Local Pain: Musings of a Mystic Scientist*. Charles Scribner's Sons, New York, 1987, p. 272.

8. p. 24. Morowitz, *Cosmic Joy and Local Pain*, p. 272.

9. p. 24-25. Morowitz, "Rediscovering The Mind," p. 16.

10. p. 25. Werner Heisenberg, *Physics and Philosophy: The Evolution in Modern Science*. Harper & Row Publishers, New York, 1958, p. 106.

11. p. 25. Paul Davies, *The Cosmic Blueprint: New Discoveries in Nature's Creative Ability to Order the Universe*. Simon & Schuster, New York, 1988, p. 165.

12. p. 27. Jane Roberts, The *God of Jane: A Psychic Manifesto*. Prentice Hall Press, New York, 1981, p. 146.

13. p. 27. Freeman Dyson, "Theology and the Origins of Life," lecture and discussion at the Center for Theology and the Natural Sciences, Berkeley, California, November 1982, p. 8.

14. p. 27. Quoted in Anthony Zee, *Fearful Symmetry: The Search for Beauty in Modern Physics*. Macmillan Publishing Company, New York, 1986, p. 280.

15. p. 27. Roberts, *The God of Jane*, p. 137.

16. p. 28. Huston Smith, *Beyond the Post-Modern Mind*. Theosophical Publishing House, Wheaton, Illinois, 1989, p. 63.

17. p. 28. Quoted in Alex Comfort, *Reality and Empathy: Physics, Mind and Science in the 21st Century*. State University of New York Press, Albany, New York, 1984, p. 24.

DAVID BOHM

1. p. 32. In physics, a particle is defined as objective when it has definite values at all times for all of its attributes.

2. p. 32. Quoted in A. P. French and Edwin F. Taylor, *An Introduction to Quantum Physics*. W. W. Norton & Company, Inc., New York, 1978, p. 278.

3. p. 32. Quoted in introduction to B. J. Hiley and F. David Peat, eds., *Quantum Implications: Essays in Honour of David Bohm*. Routledge & Kegan Paul, London and New York, 1987, p. 5.

4. p. 33. The term "underneath" refers to a subquantum level whose processes are as yet unknown to us. Such a level would undoubtedly be a region of very high energies and very small distances. See Chapter 5.

5. p. 33. Originally, physicists such as Einstein and Planck assumed that there might be a subquantum level that exhibited a continuous and causal motion. This level was thought to relate to the quantum level in a way similar to the kinetic theory of matter and the Brownian movement. At the subquantum level, there would be no quantization of energy, no wave and particle aspects, and all movement would be predictable. In short, the indeterminacy principle would not apply at this level.

6. p. 34. Fred Alan Wolf, *Parallel Universes: The Search for Other Worlds*. Simon & Schuster, 1988, p. 18.

7. p. 34. The rejection of the possibility of local hidden variable theories indicates that Heisenberg's uncertainty principle is not just a perturbation created by a measurement. Rather, Planck's constant forces us to discard the notion of well-defined mechanical parameters (e.g., position and momentum), which indicates that classical determinism is no longer universally valid.

8. p. 35. We shall be referring to the speed of light throughout this text. According to the special theory of relativity, the speed of light in a vacuum is the maximum speed attainable for a physical process. If this were not the case and a speed greater than light were attained in a given frame of reference, then the normal order of cause and effect could be reversed in another frame. The paradoxes involved are difficult to handle if causal action were to be abandoned — hence the unease that physicists feel about nonlocality.

9. p. 35. Charles Mann and Robert Crease, "Interview: John Bell." In *Omni*, Vol. 10, No. 8, May 1988, p. 90.

10. p. 35. Interview, in P. C. W. Davies and J. R. Brown, eds., *The Ghost in the Atom*. Cambridge University Press, New York, 1986, p. 48.

11. p. 36. Quoted in John Archibald Wheeler and Wojciech Hubert Zurek, eds., *Quantum Theory and Measurement*. Princeton University Press, Princeton, New Jersey, 1983, p. vi.

12. p. 36. The concept of faster-than-light propagation does not necessarily imply that a "thing" actually travels at this speed. Rather, it implies the interconnectedness of reality at a more basic level. This interconnectedness, or wholeness, contradicts the classical view of spatial separation as primary.

13. p. 36. Fred Alan Wolf, *Taking the Quantum Leap: The New Physics for Nonscientists*. Harper & Row, San Francisco, 1981, p. 177.

14. p. 37. Compared to a wave, a particle is an entity that travels in a trajectory and can be deflected and attracted, and impinge on a target at a point. These qualities describe a particle in the classical view. A wave, also a classical concept, does not travel in a trajectory, but has a property called refraction; it can vibrate and interfere with other waves.

15. p. 37. Quoted in French and Taylor, *An Introduction to Quantum Physics*, p. 54.

16. p. 38. Actually, Schrödinger originally viewed the electron as a continuous distribution of charge. Therefore, according to Schrödinger's equation, the wave became a wave of electric charge. This approach was feasible for electrons within an atom. However, when an electron is in free space, the mathematics indicated that the wave spread out into space without a limit. Therefore, the wave intensity could not represent the charge density, since the electron is always found in a very small volume of space.

17. p. 38. A complex function is composed of a real part and an imaginary part. Just as all the real numbers can be placed on a line, so can the imaginary numbers. As an example, the number 5 is real, but when it appears on the imaginary line, it becomes 5 times the square root of -1 (or i), and is written as $5i$. A complex number is located on a plane defined by a real axis and an imaginary axis. The ability to describe a system in terms of complex numbers has proven extremely advantageous to the physicist. But, unfortunately, we never see complex numbers directly. That is, complex numbers seem to underlie our three-dimensional world, but we can never measure them because they bear no direct relationship with physical reality as real numbers do.

18. p. 38. The observable quantities are obtained by multiplying the complex wave function by its complex conjugate (explained below). The product is always a real number. Furthermore, a physicist would define a "real" field as having energy at space-time points remote from the presence of matter. For those readers who are interested, the wave function utilizes complex functions of the form $x+iy$. To obtain the probability in question, the amplitude of the wave function must be squared. To make sure that this process yields a real number, the amplitude (complex function) is multiplied by what is termed its complex conjugate. If $x+iy$ is the amplitude, then $x-iy$ is the complex conjugate.

19. p. 40. David Bohm and F. David Peat, *Science, Order, and Creativity*. Bantam Books, New York, 1987, p. 40.

20. p. 41. John Archibald Wheeler, "Law Without Law," in *Quantum Theory and Measurement*, p. 194.

21. p. 42. Seth proposes a similar concept (discussed in Chapter 3): "There are no real divisions between the perceiver and the thing seemingly perceived. In many ways the thing perceived is an extension of the perceiver." Jane Roberts, *Seth Speaks: The Eternal Validity of the Soul*. Prentice Hall Press, New York, 1987, p. 94.

22. p. 42. Remember that to assume that a quantum particle is a distinct entity in the classical sense requires the subject-object split.

23. p. 43. Bohm and Peat, *Science, Order, and Creativity*, pp. 6-7.

24. p. 43-44. Albert Einstein and Leopold Infeld, *The Evolution of Physics*. Simon & Schuster, New York, 1938, p. 291.

25. p. 44. Quoted in John D. Barrow, *The World Within the World*. Oxford University Press, New York, 1988, p. 279. As an example, mathematics cannot represent the beauty of a rose or the sweetness of a peach. Information, to have meaning, must be inserted into a system of facts. The system can enhance comprehension or hinder it. The expression of human experience requires numerous systems, of which mathematics is one.

26. p. 45. Nick Herbert, *Quantum Reality: Beyond the New Physics*. Anchor Press/ Doubleday, Garden City, New York, 1985, p. 40.

27. p. 46. Barrow, *The World Within the World*, p. 193.

28. p. 47. David Bohm and B. J. Hiley, "On the Intuitive Understanding of Nonlocality as Implied by Quantum Theory," in *Foundations of Physics*, Vol. 5, No. 1, 1975, p. 102.

29. p. 47. Rudy Rucker, *Mind Tools: The Five Levels of Mathematical Reality*. Houghton Mifflin Company, Boston, 1987, p. 188.

30. p. 48. Supergravity theories have proposed up to eleven dimensions, with seven dimensions rolled up. A superstring theory postulates ten dimensions.

31. p. 48. Bohm applies this concept to particles:

> Quantum mechanics implies that . . . two three-dimensional particles are not independently existent elements, but rather, they are each a three-dimensional projection of a higher six-dimensional reality. With n particles, each is a three-dimensional projection of a $3n$ dimensional reality. We cannot picture this situation, but . . . both experimentally and theoretically, quantum mechanics strongly indicates the need for this new view of the higher-dimensional reality, which is the ground of our ordinary three-dimensional reality.

"Issues in Physics, Psychology and Metaphysics: A Conversation," in *Journal of Transpersonal Psychology*, Vol. 12, No. 1, 1980, p. 28.

32. p. 49. Mann and Crease, "Interview: John Bell," p. 90.

33. p. 50. Interview in Davies and Brown, eds., *The Ghost in the Atom*, p. 140.

34. p. 50. According to the Heisenberg uncertainty principle, it is not possible to accurately measure both the position and momentum of a particle at the same time. There is a mathematical limit to the accuracy such that the more accurately you measure the position of a particle, the less accurately you can measure the momentum. If the position was precisely known, the momentum would be completely uncertain, and if the momentum were precisely known, the position would be completely uncertain.

35. p. 52. Renée Weber, *Dialogues with Scientists and Sages: The Search for Unity*. Routledge & Kegan Paul, London and New York, 1986, p. 235.

36. p. 52. Ilya Prigogine, *From Being to Becoming: Time and Complexity in the Physical Sciences*. W. H. Freeman and Company, New York, 1980, p. xiii.

37. p. 52. Quoted in Jefferson Hane Weaver, *The World of Physics,* Vol. 2. Simon & Schuster, New York, 1987, p. 699.

38. p. 52. Chris Isham, "Quantum Gravity," in Paul Davies, ed., *The New Physics*. Cambridge University Press, New York, 1989, pp. 71-72.

39. p. 53. Jeremy Campbell, *Grammatical Man: Information, Entropy, Language, and Life*. Simon & Schuster, Inc., New York, 1982, p. 16.

40. p. 60. Bohm and Hiley, "On The Intuitive Understanding of Nonlocality as Implied by Quantum Theory," p. 104.

41. p. 62. James Gleick, *Chaos: Making A New Science*. Viking Penguin Inc., New York, 1987, jacket.

42. p. 62. Jane Roberts, *The Nature of the Psyche: Its Human Expression*. Bantam Books, New York, 1979, p. 170.

43. p. 64. Weber, *Dialogues with Scientists and Sages,* p. 26.

44. p. 65. Quoted in interview, Davies and Brown, eds., *The Ghost in the Atom*, p. 66.

45. p. 65. The quantum field theory partially unifies special relativity theory and quantum mechanics. It is a mathematical formalism that applies to fields. As we have indicated previously, a field varies from point to point. By this very nature, fields have wave properties. The application of quantum mechanics to a field results in a structure that can lead to a particle interpretation and a better understanding of the duality of wave and particle. The most successful theory of this kind is known as quantum electrodynamics. However, even this remarkably accurate formalism still gives nonsensical infinite answers that must be removed by a process called renormalization. Renormalization allows the theory to issue accurate predictions, but its mathematical consistency is open to question. Because of that problem, some physicists feel that a deeper understanding is required.

46. p. 66. Weber, *Dialogues with Scientists and Sages,* p. 27.

47. p. 67. Quoted in Weaver, *The World of Physics*, Vol. 2, p. 431.

48. p. 68. Quoted in Weaver, *The World of Physics*, Vol. 3, p. 681.

49. p. 69. Donald Factor, ed., *Unfolding Meaning: A Weekend of Dialogue with David Bohm*. Foundation House Publications, Glouchestershire, England, 1985, p. 14.

50. p. 70. F. David Peat, *Synchronicity: The Bridge Between Matter and Mind*. Bantam Books, New York, 1987, pp. 87-88.

51. p. 72. Quoted in William Irwin Thompson, ed., *Gaia: A Way of Knowing.* Lindisfarne Press, Great Barrington, Massachusetts, 1987, p. 27.

52. p. 72. Roberts, *Seth Speaks,* p. 406.

53. p. 73. Physicists can reverse the direction of time by making $t = -t$. Most of the equations of physics are symmetrical with respect to time; that is, they work the same whether one is going into the past or the future. The reversibility of time applies not only to Newton's equations but also to Maxwell's equations and to general relativity theory. Even Schrödinger's equation is time-symmetrical before the wave function collapses.

54. p. 74. For readers familiar with the paradox of Schrödinger's cat (described on pp. 249-250), the symmetry is broken when the physicist looks at the condition of the cat. It is alive or dead at that point, not both. Using Bohm's terminology, at the point of observation, the cat is explicated, dead or alive. However, with Bohm the superquantum potential of the cat can cause the explication without the need for the presence of the physicist. This is treated in more detail in Chapter 7.

55. p. 78-79. Quoted in Robert G. Jahn and Brenda J. Dunne, *Margins of Reality: The Role of Consciousness in the Physical World.* Harcourt Brace Jovanovich, Orlando, 1987, p. 267.

56. p. 81. Renée Weber, "Meaning as Being in the Implicate Order Philosophy of David Bohm: A Conversation," in *Quantum Implications: Essays in Honour of David Bohm*, p. 436.

57. p. 85. Heinz R. Pagels, *The Dreams of Reason: The Computer and the Rise of the Sciences of Complexity.* Simon & Schuster, New York, 1988, p. 211.

58. p. 88. Alex Comfort, *Reality and Empathy: Physics, Mind, and Science in the 21st Century.* State University of New York Press, Albany, New York, 1984, p. 22.

59. p. 88. Renée Weber, "The Physicist and the Mystic — Is a Dialogue Between Them Possible? A Conversation with David Bohm," in Ken Wilber, ed., *The Holographic Paradigm and Other Paradoxes.* Shambhala Publications, Boulder, Colorado, 1982, p. 197.

60. p. 89. When two observables do not commute, it indicates that if they were multiplied, the order of the observables will lead to different results. As an example, A X B will not equal B X A if A and B do not commute.

61. p. 89-90. David Bohm, *Wholeness and the Implicate Order.* Routledge & Kegan Paul, London, 1983, p. 176.

62. p. 90. Quoted in David Bohm and B. J. Hiley, "Nonlocality in Quantum Theory Understood in Terms of Einstein's Nonlinear Field Approach," in *Foundations of Physics*, Vol. 11, Nos. 7/8, 1981, p. 533.

63. p. 90. Interview in Davies and Brown, eds., *The Ghost in the Atom*, p. 56.

64. p. 91. Interview in Davies and Brown, eds., *The Ghost in the Atom*, p. 142.

THE PERENNIAL PHILOSOPHY

1. p. 96. Quoted in Henry Margenau, *The Miracle of Existence*. Shambhala Publications, Boston, pp. 114-115.

2. p. 97. Aldous Huxley, *The Perennial Philosophy*. Harper & Row Publishers, New York, p. 35.

3. p. 97. In the Perennial Philosophy and in Jungian writing, "self" is often capitalized when it refers to the "higher" aspect of the self. Because this usage does not correspond to that of other sources considered here, we have chosen not to use the capital letter.

4. p. 97. Wilber has written books on sociology (*A Sociable God*), evolution (*Up From Eden*), consciousness (*The Spectrum of Consciousness*), human development (*The Atman Project*), paradigm development (*Eye to Eye*), and other topics, and is editor-in-chief of *Revision Journal* and general editor of New Science Library.

5. p. 98. See discussion of Hamilton-Jacobi theory, pp. 39-41.

6. p. 99. Ken Wilber, ed., *The Holographic Paradigm and Other Paradoxes*. Shambhala Publications, Boulder, Colorado, 1982, p. 160.

7. p. 100. Wilber, ed., *The Holographic Paradigm and Other Paradoxes*, p. 159.

8. p. 101. Recall from our discussion of the big bang theory (pp. 85-87) that the void before the big bang is literally unimaginable but fertile with physical law. In this sense, Wilber's deep structure and the pre-existing void are analogous.

9. p. 103. Ken Wilber, *The Atman Project: A Transpersonal View of Human Development*. Theosophical Publishing House, Wheaton, Illinois, 1980, p. 80.

10. p. 104. Wilber, *The Atman Project*, p. 89.

11. p. 104. Wilber, *The Atman Project*, p. 42.

12. p. 106. Jane Roberts, *Seth Speaks: The Eternal Validity of the Soul*. Prentice Hall Press, New York, 1972, p. 303-304.

13. p. 107. Wilber, ed., *The Holographic Paradigm and Other Paradoxes*, p. 161.

14. p. 110. Wilber, *The Atman Project*, p. 175.

15. p. 111. Roberts, *Seth Speaks*, p. 15.

16. p. 112. Ken Wilber, *Eye to Eye: The Quest for the New Paradigm*. Anchor Press/Doubleday, Garden City, New York, 1983, p. 298.

17. p. 112. Wilber, *Eye to Eye*, p. 297.

SETH

1. p. 115. It is important to acknowledge that even in our three-dimensional world, different consciousnesses perceive the world in different ways. Certainly a frog does not see the world as a human being does. Entire civilizations differ in their perceptions of reality, and each individual perceives the world based on his or her own propensities and experience. It is as if each consciousness takes the world and molds it to its own predilections.

2. p. 116. Quoted in Jane Roberts, *The Seth Material*. Prentice Hall, Inc., New York, 1970, p. 111.

3. p. 117. Marie-Louise von Franz, *On Dreams and Death: A Jungian Interpretation*. Shambhala Publications, Boston, 1986, p. 153.

4. p. 119. Jane Roberts, *Seth Speaks: The Eternal Validity of the Soul*. Prentice Hall Press, New York, 1972. p. 10.

5. p. 120. Roberts, *Seth Speaks*, p. 10.

6. p. 120. Ernest R. Hilgard and Josephine R. Hilgard, *Hypnosis in the Relief of Pain*. William Kaufmann, Inc., Los Altos, California, 1983. The experiment discussed here is described on page 167.

7. p. 121. Hilgard and Hilgard, *Hypnosis in the Relief of Pain*, p. 230.

8. p. 121. Jeremy Campbell, *Grammatical Man: Information, Entropy, Language, and Life*. Simon & Schuster, Inc., New York, 1982, p. 237.

9. p. 122. Quoted in Renée Weber, *Dialogues with Scientists and Sages: The Search for Unity*. Routledge & Kegan Paul, London and New York, 1986, p. 135.

10. p. 122. Roger S. Jones, *Physics as Metaphor*. New American Library, New York, 1982, p. 3.

11. p. 123. Roberts, *The Seth Material*, p. 137.

12. p. 124. Jane Roberts, *The "Unknown" Reality*. Vol. 1, Prentice Hall Press, New York, 1986, p. 66.

13. p. 124. Weber, *Dialogues with Scientists and Sages*, p. 235.

14. p. 124. Robert G. Jahn and Brenda J. Dunne, *Margins of Reality: The Role of Consciousness in the Physical World*. Harcourt Brace Jovanovich, Orlando, 1987, p. 61.

15. p. 124-125. Jane Roberts, *The Nature of the Psyche: Its Human Expression*. Bantam Books, New York, 1979, p. 163.

16. p. 125. Jane Roberts, *Dreams, "Evolution," and Value Fulfillment*. Vol. 1. Prentice Hall Press, New York, 1986, p. 170.

17. p. 126. Weber, *Dialogues with Scientists and Sages*, p. 45.

18. p. 127. Roberts, *Seth Speaks*, p. 75.

19. p. 128. Roberts, *Seth Speaks*, p. 77.

20. p. 128. Bob Toben and Fred Alan Wolf, *Space-Time and Beyond: Toward an Explanation of the Unexplainable*, new ed. Bantam Books, Inc., New York, 1983, p. 130.

21. p. 128. Ken Wilber, ed., *The Holographic Paradigm and Other Paradoxes*. Shambhala Publications, Boulder, Colorado, 1982, p. 198.

22. p. 129. Quoted in Jahn and Dunne, *Margins of Reality*, p. 247.

23. p. 129. B. J. Hiley and F. David Peat, eds., *Quantum Implications: Essays in Honour of David Bohm*. Routledge & Kegan Paul, London and New York, 1987, p. 443.

24. p. 129. Hiley and Peat, eds., *Quantum Implications*, p. 443.

25. p. 130. David Bohm and F. David Peat, *Science, Order, and Creativity*. Bantam Books, New York, 1987, p. 93.

26. p. 130. Bohm and Peat, *Science, Order, and Creativity*, p. 93.

27. p. 131. Gary Zukav, *The Dancing Wu Li Masters: An Overview of the New Physics*. William Morrow and Company, Inc., New York, 1979, p. 327.

28. p. 131. Jahn and Dunne, *Margins of Reality*, p. 240.

29. p. 132. See notes 17 and 18, Chapter 1.

30. p. 133. Roberts, *Seth Speaks*, p. 360.

31. p. 135. Roberts, *Seth Speaks*, p. 278.

32. p. 135. Roberts, *The Nature of the Psyche*, p. 165.

33. p. 136. Ron Atkin, *Multidimensional Man*. Penguin Books, Harmondsworth, England, 1981, p. 173.

34. p. 136. Atkin, *Multidimensional Man*, p. 174.

35. p. 137. Nick Herbert, "Scientists Explore Invisible Ocean of Glue." C-Life Institute, Boulder Creek, California, 1977, p. 2.

36. p. 137. Weber, *Dialogues with Scientists and Sages*, p. 93.

37. p. 138. Roberts, *Seth Speaks*, 279.

38. p. 138-139. Roberts, *The Seth Material*, p. 126.

39. p. 140. John Archibald Wheeler, "Law Without Law," in John Archibald Wheeler and Wojciech Hubert Zurek, eds., *Quantum Theory and Measurement*. Princeton University Press, Princeton, New Jersey, 1983, p. 190.

40. p. 141. John D. Barrow, *The World Within The World*. Oxford University Press, Oxford, 1988, pp. 153-154.

41. p. 141. Jahn and Dunne, *Margins of Reality,* p. 321.

42. p. 142. Wheeler, "Law Without Law," p. 194.

43. p. 142. William Irwin Thompson, *Pacific Shift*. Sierra Club Books, San Francisco, 1985, p. 65.

44. p. 143. Roberts, *The Nature of the Psyche*, p. 227.

45. p. 143. Roberts, *The Nature of the Psyche*, p. 227.

46. p. 144. Alex Comfort, *Reality and Empathy: Physics, Mind, and Science in the 21st Century*. State University of New York Press, Albany, New York, 1984, p. 35.

47. p. 144. David Bohm, *Wholeness and the Implicate Order*. Routledge & Kegan Paul, London, 1983, p. 192.

48. p. 145. Bohm, *Wholeness and the Implicate Order*, p. 192.

49. p. 145. Jane Roberts, *The Individual and the Nature of Mass Events*. Prentice Hall Press, New York, 1987, p. 94.

50. p. 147. Roberts, *The Individual and the Nature of Mass Events*, pp. 80-81.

51. p. 147-148. Roberts, *The Individual and the Nature of the Mass Events*, pp. 82-83.

52. p. 149. Roberts, *The Individual and the Nature of Mass Events*, p. 90.

53. p. 151-152. Roberts, *Seth Speaks*, p. 431.

54. p. 152. Roberts, *The Individual and the Nature of Mass Events*, pp. 109-110.

55. p. 152. Jane Roberts, *The "Unknown" Reality*, Vol. 1. Prentice Hall Press, New York, 1986, p. 69.

56. p. 154. Campbell, *Grammatical Man*, p. 237.

57. p. 155. Roberts, *The Nature of the Psyche*, p. 217.

58. p. 156. Jane Roberts, *The "Unknown" Reality*, Vol. 2. Prentice Hall Press, New York, 1986, p. 458.

59. p. 156. Roberts, *The "Unknown" Reality*, Vol. 2. p. 459.

60. p. 158. In *The 3-Pound Universe* (Macmillan Publishing Company, New York, 1986, pp. 281-282), Judith Hooper and Dick Teresi relate this story about lucid dreaming. LaBerge's star subject for lucid dream experiments was a young woman, Beverly Kedzierski, who had her first lucid dream when she was five years old. In a recurrent nightmare, witches chased her around the yard. Each time she had the dream, Beverly bargained with the witches to let her go, promising to let them catch her the next time. Each time, the witches acquiesced, then chased her again the next time. Finally, Beverly became exasperated and in her dream demanded to know why the witches were chasing her. They did not answer, and she told them to go away. They left and never returned. The witches, Seth would say, were the shadows or hallucinations of Beverly's own thoughts and emotions.

61. p. 158. Roberts, *Dreams, "Evolution," and Value Fulfillment*, Vol. 2, p. 462.

62. p. 159. F. David Peat, *Synchronicity: The Bridge Between Matter and Mind*. Bantam Books, 1987, pp. 123-124.

63. p. 160. Roberts, *The "Unknown" Reality*, Vol. 1, p. 70.

64. p. 161-162. Fred Alan Wolf, *Star Wave: Mind, Consciousness, and Quantum Physics*. Macmillan Publishing Co., New York, 1984, pp. 187-188.

65. p. 162. Nick Herbert, *Faster than Light: Superluminal Loopholes In Physics*. Penguin, New York, 1988, p. 39.

COMMON ELEMENTS

1. p. 167. The concept of levels was mythologically represented as far back as the ancient Greeks. The earth was conceived as a stationary sphere centered within a set of transparent revolving spheres, each of which was associated with a heavenly body (sun, moon, or planet) and with a metal (silver, mercury, copper, etc.). The soul descended from an area outside the largest sphere (where God presided) down through the spheres, finally to be born on earth. In its descent it picked up the metals needed to constitute a body here on earth. At death, the soul returned, divested of these metals en route, and arrived in its original state in the presence of God. This myth suggests a hierarchy similar to the ones discussed here, especially in Wilber's concepts of involution and the Atman-project.

2. p. 169. Evan Walker, "Foundations of Paraphysical and Parapsychological Phenomena," in Laura Oteri, ed., *Quantum Physics and Parapsychology*, Proceedings of an International Conference held in Geneva, Switzerland, August 26-27, 1974, Parapsychology Foundation, Inc., New York.

3. p. 169. Ken Wilber, *The Spectrum of Consciousness*. Theosophical Publishing House, Wheaton, Illinois, 1977, p. 16.

4. p. 171. Renée Weber, *Dialogues with Scientists and Sages: The Search for Unity*. Routledge & Kegan Paul, London and New York, 1986, p. 28.

5. p. 171-172. Weber, *Dialogues with Scientists and Sages*, p. 151.

6. p. 172. Ken Wilber, ed., *The Holographic Paradigm and Other Paradoxes*. Shambhala Publications, Boulder, Colorado, 1982, p. 169.

7. p. 172. Ken Wilber, *The Atman Project: A Transpersonal View of Human Development*. Theosophical Publishing House, Wheaton, Illinois, 1980, p. 83.

8. p. 172. Wilber,*The Atman Project*, p. 80.

9. p. 173. Weber, *Dialogues with Scientists and Sages*, p. 40.

10. p. 173. Weber, *Dialogues with Scientists and Sages,* p. 93.

11. p. 174. Wilber, *The Atman Project*, p. 80.

12. p. 175. Donald Factor, ed., *Unfolding Meaning: A Weekend of Dialogue with David Bohm.* Foundation House Publications, England, 1985, p. 102. This book is based on a workshop held in England in 1984 to inquire more deeply into some of Bohm's thoughts on a variety of subjects. The excerpts that follow are from pages 102-107.

13. p. 176. B. J. Hiley and F. David Peat, eds., *Quantum Implications: Essays in Honour of David Bohm*. Routledge & Kegan Paul, London and New York, 1987, p. 436.

14. p. 177. Hiley and Peat, eds., *Quantum Implications*, p. 438.

15. p. 177. Wilber, ed., *The Holographic Paradigm and Other Paradoxes*, p. 214.

16. p. 177. Jane Roberts, *The Nature of the Psyche: Its Human Expression*. Prentice Hall Press, New York, 1979, p. 222.

17. p. 178. Rudy Rucker, *Infinity and the Mind: The Science and Philosophy of the Infinite*. Bantam Books, Inc., New York, 1983, p. 181.

18. p. 178-179. Weber, *Dialogues with Scientists and Sages*, p. 94.

19. p. 179-180. Roberts, *The Nature of the Psyche*, p. 170.

20. p. 180. "Kurt Gödel," in *Encyclopedia Britannica*, (Micropedia), Vol. 5, Chicago, 1986.

21. p. 181. Quoted in Edna E. Kramer, *The Nature and Growth of Modern Mathematics*. Princeton University Press, Princeton, New Jersey, 1981, p. 577.

22. p. 182-183. Wilber, *The Atman Project*, p. 38-39.

23. p. 183. Wilber, *The Spectrum of Consciousness*, p. 91.

24. p. 184. David Bohm, *Wholeness and the Implicate Order*. Routledge & Kegan Paul, London, 1983, p. 77.

25. p. 185. Jane Roberts, *Seth Speaks: The Eternal Validity of the Soul*. Prentice Hall Press, New York, 1987, p. 362.

26. p. 185. Jane Roberts, *The God of Jane: A Psychic Manifesto*. Prentice Hall Press, New York, 1987, p. 14.

27. p. 186. Jane Roberts, *The Individual and the Nature of Mass Events*. Prentice Hall Press, New York, 1987, p. 93.

28. p. 186. Roberts, *The God of Jane*, p. 139.

29. p. 186-187. Jane Roberts, *Dreams and Projection of Consciousness*. Stillpoint Publishing, Walpole, New Hampshire, p. 314.

30. p. 187. Roberts, *The God of Jane*, p. 14.

31. p. 187. Wilber, *The Atman Project*, p. 40.

32. p. 187-188. Jane Roberts, *Dreams, "Evolution," and Value Fulfillment*, Vol. 1. Prentice Hall Press, New York, 1986, p. 256.

33. p. 188. Quoted in Kenneth Ring, *Life at Death: A Scientific Investigation of the Near-Death Experience*. Coward, McCann & Geoghegan, New York, 1980, p. 238.

34. p. 188. Roberts, *The God of Jane*, p. 139.

35. p. 189. Roberts, *Seth Speaks*, p. 12.

36. p. 190. Quoted in Gerald Holton and Yehuda Elkana, eds., *Albert Einstein, Historical and Cultural Perspectives*. Princeton University Press, Princeton, New Jersey, 1982, p. 69.

37. p. 193. Roberts, *Seth Speaks*, p. 372.

SPACE-TIME CREATION

1. p. 204. Bob Toben and Fred Alan Wolf, *Space-Time and Beyond: Toward an Explanation of the Unexplainable*, new ed. E. P. Dutton, New York, 1982, p. 146.

2. p. 204. Quoted in Bob Toben, *Space-Time and Beyond*. E. P. Dutton, New York, 1975, p. 127.

3. p. 206-207. Richard P. Feynman, *QED: The Strange Theory of Light and Matter*. Princeton University Press, Princeton, New Jersey, 1985, p. 128.

4. p. 207. David Bohm, *Wholeness and the Implicate Order*. Routledge & Kegan Paul, London and Boston, 1983, p. 83.

5. p. 209. Jane Roberts, *Dreams, "Evolution," and Value Fulfillment*, Vol.1. Prentice Hall Press, New York, 1986, p. 137.

6. p. 209-210. Jane Roberts, *The "Unknown" Reality*, Vol. 1. Prentice Hall Press, New York, 1986, pp. 125, 126.

7. p. 210. Jane Roberts, *Seth Speaks: The Eternal Validity of the Soul*. Prentice Hall Press, New York, 1987,p. 453.

8. p. 211. Quoted in Toben, *Space-Time and Beyond*, p. 127.

9. p. 213. Roberts, *Seth Speaks*, p. 360.

10. p. 213. Jane Roberts, *The "Unknown" Reality*, Vol. 2. Prentice Hall Press, New York, 1986, p. 328.

11. p. 213. Renée Weber, *Dialogues with Scientists and Sages: The Search for Unity*. Routledge & Kegan Paul, New York and London, 1986, p. 40.

12. p. 215. Marie-Louise von Franz, *On Dreams and Death*. Shambhala Publications, Boston, 1986, p. 147.

13. p. 216. Quoted in P. C. W. Davies and J. Brown, eds., *Superstrings: A Theory of Everything?* Cambridge University Press, Cambridge, England, 1988, pp. 124-125.

14. p. 216. Quoted in Davies and Brown, eds., *Superstrings*, p. 119.

15. p. 217. Quoted in Davies and Brown, eds., *Superstrings*, p. 123.

16. p. 218. F. David Peat, *Superstrings and the Search for the Theory of Everything*. Contemporary Books, Chicago, 1988, pp. 12-13.

17. p. 218. F. David Peat, *Synchronicity: The Bridge Between Matter and Mind*. Bantam Books, 1987, p. 183.

18. p. 218-219. Paul Davies, *Superforce: The Search for a Grand Unified Theory of Nature*. Simon & Schuster, New York, 1984, p. 109.

19. p. 219. Quoted in Jefferson Hane Weaver, *The World of Physics*, Vol. 3. Simon & Schuster, New York, 1987, p. 681.

20. p. 221. Edward R. Harrison, *Cosmology: The Science of the Universe*. Cambridge University Press, Cambridge, England, 1981, pp. 191-192.

21. p. 222. Lawrence Leshan and Henry Margenau, *Einstein's Space and Van Gogh's Sky: Physical Reality and Beyond.* Macmillan Publishing Company, New York, 1982, p. 110.

22. p. 222. Roberts, *The "Unknown" Reality*, Vol. 1, p. 229.

23. p. 222. A study made by Wolfgang Pauli in 1925 showed that the interaction of the wave properties of particles is such that two particles cannot occupy the same quantum state. Mathematically, two electrons exhibit a kind of repulsion when brought close together, which is in addition to any electrical forces that are present.

24. p. 223. The angular momentum of a ball is normally the product of the moment of inertia times the angular velocity. If the ball is an isolated system, the angular momentum will always be constant (analogous to the conservation of momentum). But the moment of inertia of a ball is a function of its radius. If the radius decreases, the moment of inertia decreases. In order to conserve angular momentum, as the radius decreases, the angular velocity increases. If an electron is a point, the angular velocity is infinite. Hence, electron spin is not similar to a ball spinning. According to the Dirac equation, the spin is the motion of the cluster of probable locations around the point of maximum probability. Such motion is larger than the envisioned size of an electron, and therefore its spin is reasonable, being less than the velocity of light.

25. p. 223. Davies, *Superforce*, p. 34.

26. p. 224. Dirac's equation led to the merging of quantum theory and Einstein's special theory of relativity.

    Schrödinger's equation described the motion of electrons in terms of waves but did not incorporate the concept of relative motion and the relationship (Einsteinian) between mass and energy, which Dirac's equation did. But Dirac came up with Einstein's famous equation $E=mc^2$ in a slightly different form, $E^2=m^2c^4$. If the square root of both sides is taken, then there is also the root $E=-mc^2$. The minus sign led to antimatter.

27. p. 224. Heinz R. Pagels, *Perfect Symmetry: The Search for the Beginning of Time*. Simon & Schuster, New York, 1985, pp. 191-192.

28. p. 224. Pagels, *Perfect Symmetry*, pp. 192-193.

29. p. 225. Roberts, *Seth Speaks*, p. 490.

30. p. 226. Roberts, *Seth Speaks*, p. 362.

31. p. 226. John Gribbin, *In Search of Schrödinger's Cat: Quantum Physics and Reality*. Bantam Books, New York, 1984, p. 187.

32. p. 226. Roberts, *Seth Speaks*, p. 491.

33. p. 226. Quoted in Toben, *Space-Time and Beyond*, p. 127.

34. p. 228-229. von Franz, *On Dreams and Death*, p. 155.

35. p. 230. Roberts, *Seth Speaks*, p. 139.

36. p. 232. Jane Roberts, *Dreams and Projections of Consciousness*. Stillpoint Publishing, Walpole, New Hampshire, 1987, p. 47.

37. p. 232. See the books of Carlos Castaneda, beginning with *The Teachings of Don Juan: A Yaqui Way of Knowledge* (1968, copyright by the Regents of the University of California, Ballantine Books, New York).

THE MIND-BODY PROBLEM

1. p. 234. David Bohm and F. David Peat, *Science, Order and Creativity*. Bantam Books, New York, 1987, page 101.

2. p. 234. Nick Herbert, "Consciousness As A Problem In Physics," C-Life Institute, Boulder Creek, California, 1977, p. 3.

3. p. 234-235. Jane Roberts, *The Seth Material*. Prentice Hall Press, New York, 1970, pp. 125-126.

4. p. 237. F. David Peat, *Synchronicity: The Bridge Between Mind and Matter*. Bantam Books, New York, 1987, pp. 63-64.

5. p. 239. B. J. Hiley and F. David Peat, eds., *Quantum Implications: Essays in Honour of David Bohm*. Routledge & Kegan Paul, London and New York, 1987, p. 38.

6. p. 239. Jane Roberts, *Seth Speaks: The Eternal Validity of the Soul*. Prentice Hall Press, New York, 1987, p. 12.

7. p. 239. Paul Davies, *The Cosmic Blueprint*. Simon & Schuster, New York, 1988, p. 149.

8. p. 240. Jane Roberts, *The "Unknown" Reality*, Vol. 1. Prentice Hall Press, New York, 1986, page 235.

9. p. 240. Roberts, *The Seth Material*, p. 125.

10. p. 240. Jane Roberts, *The "Unknown" Reality*, Vol. 2. Prentice Hall Press, New York, 1986, p. 323.

11. p. 240. *Brain / Mind Bulletin*, Los Angeles, November 1987, Vol. 13, No. 2, p. 1.

12. p. 241. In classical physics, if equations are transferred from one coordinate system to another, the laws of mechanics remained the same, or invariant. This is the classical transformation law. Relativity theory required a new transformation law, and it is called the Lorentz transformation. All laws of physics must now be invariant with respect to the Lorentz transformation. Bell is saying here that if you devise laws for the interaction of mind and body, they must be Lorentz-invariant.

13. p. 241. P. C. W. Davies and J. R. Brown, eds., *The Ghost in the Atom*. Cambridge University Press, New York, 1986, p. 54.

THE MEASUREMENT PROBLEM

1. p. 244. John D. Barrow, *The World Within the World*. Oxford University Press, New York, 1988, pp. 133-136.

2. p. 251. Bohm explains the concept of a three-dimensional projection of a higher-dimensional reality. Refer to Chapter 1, Classical Potentials and the Quantum Potential (pp. 50-56.)

3. p. 252. Jane Roberts, *The Seth Material*. Prentice Hall Press, New York, 1970, p. 127.

BELL'S THEOREM

1. p. 255. Quoted in John Archibald Wheeler and Wojciech Hubert Zurek, eds., *Quantum Theory and Measurement*. Princeton University Press, Princeton, New Jersey, 1983, p. 8.

2. p. 257. In assigning objective reality to particle B only after a measurement, Bohr left the door open for injecting the physicist's mind into the laws of physics. Nobel-prizewinning physicist Eugene Wigner walked right in. Wigner has been quoted as follows:

   > [I]t was not possible to formulate the laws of quantum mechanics in a fully consistent way without reference to the consciousness [of the observer] . . . the very study of the external world led to the conclusion that the content of the consciousness is the ultimate reality.

   Robert P. Crease and Charles C. Mann, *The Second Creation*. MacMillan Publishing Co., New York, 1986, p. 67.

3. p. 257. Heinz R. Pagels, *The Cosmic Code: Quantum Physics as the Language of Nature*. Bantam Books, 1983, p. 76.

4. p. 258. Conseil Europeen pour la Recherche Nucleaire.

5. p. 258-259. Nick Herbert, *Quantum Reality: Beyond the New Physics*. Anchor Press/Doubleday, Garden City, New York, 1985, pp. 214-215.

6. p. 260. In D. J. Bohm and B. J. Hiley, "On The Intuitive Understanding of Nonlocality as Implied by Quantum Theory," *Foundations of Physics*, Vol. 5, No. 1, 1975, pp. 106-107.

7. p. 261. Quoted in interview in P. C. W. Davies and J. R. Brown, eds., *The Ghost in the Atom*. Cambridge University Press, Cambridge, England, 1986, p. 143.

8. p. 261. Quoted in Jeffrey Mishlove, *The Roots of Consciousness: Psychic Liberation Through History, Science, and Experience*. A Random House, New York, 1975, p. 279.

DISSIPATIVE SYSTEMS

1. p. 266-267. Morris Kline, *Mathematics and the Search for Knowledge*. Oxford University Press, New York, 1985, pp. 239-240.

2. p. 268. Louise B. Young, *The Unfinished Universe*. Simon & Schuster, New York, 1986, pp. 15-16.

EASTERN AND WESTERN THOUGHT

1. p. 271. Diffraction is a phenomenon associated with all forms of waves. If an obstacle comparable in size to the wavelength is placed in the path of a wave, some parts of the wave bend around it into the region of the "shadow" produced by the obstacle so that the shadow has edges that are less sharp than if the radiation consisted of particles. As a result, in some parts of the nonshadow region, the intensity of the radiation is reduced.

2. p. 272. Jane Roberts, *Dreams, "Evolution," and Value Fulfillment*, Vol. 1. Prentice Hall Press, New York, 1986, p. 169.

3. p. 275. Henry Margenau, *The Miracle of Existence*. Shambhala Publications, Inc., Boston, 1987, p. 115.

4. p. 275. Fritjof Capra, *Uncommon Wisdom: Conversations with Remarkable People*. Simon & Schuster, New York, 1988, p. 40.

5. p. 275. Jane Roberts, *The "Unknown" Reality*, Vol. 1. Prentice Hall Press, New York, 1986, p. 190.

6. p. 276. Roberts, *The "Unknown" Reality*, Vol. 1, pp. 190-191.

7. p. 276. William Irwin Thompson, *Pacific Shift*. Sierra Club Books, San Francisco, 1985, p. 18.

8. p. 276-277. Roberts, *The "Unknown" Reality*, Vol. 1, pp. 205-206.

9. p. 277-278. Roberts, *The "Unknown" Reality*, Vol. 1, p. 228.

10. p. 279. Fred Alan Wolf, *Star Wave: Mind, Consciousness, and Quantum Physics*. Macmillan Publishing Company, New York, 1984, pp. 232 and 234.

11. p. 279-280. Wolf, *Star Wave*, p. 235.

TOWARD A NEW PARADIGM

1. p. 281. This statement reflects Hawking's basic agreement with the Newtonian model of reality. Apparently, at the time of this lecture, Hawking felt that a supergravity theory would lead to a single formula (or set of formulae) from which the entire universe could be explained. Supergravity theories attempt to unify gravitational (general relativity) theory with the theories of electromagnetic, weak, and strong interactions.

2. p. 285. David Bohm and F. David Peat, *Science, Order, and Creativity*. Bantam Books, New York, 1987, p. 80.

3. p. 285. Bohm and Peat, *Science, Order, and Creativity*, p. 81.

4. p. 286. Jane Roberts, *Dreams, "Evolution," and Value Fulfillment*, Vol. 1. Prentice Hall Press, New York, 1986, p. 114.

5. p. 291. Jane Roberts, *The Seth Material*. Prentice Hall Press, New York, 1970, p. 137.

6. p. 291. Renée Weber, *Dialogues with Scientists and Sages: The Search for Unity*. Routledge & Kegan Paul, London and New York, 1986, p. 19.

Glossary

algorithm A calculational procedure containing a well-defined sequence of operations.

All That Is A term from the Seth material meaning the vast and infinite psychological realm whose energy is within and behind all formations. It is alive within the least of Itself. All of Its creations are endowed with Its own abilities, and, it is infinitely becoming, never complete.

antiparticle The counterpart of an elementary particle with identical mass and spin and opposite electric charge and magnetic properties. When a particle and its antiparticle collide, they are destroyed, with the equivalent energy surviving as radiation. Some electrically neutral particles are their own antiparticles. Fortunately, antiparticles are usually seen only in laboratories; the rare exception is in the debris of a cosmic ray shower.

archetype A predisposition toward certain patterns of psychological performance that are passed down from our ancestral past (human and animal) and are linked to instinct. In Platonic terms, an archetype is the underlying form or idea of a thing. According to Jung, archetypes are forms without content in the collective unconscious, which become activated as psychic force.

Atman (Sanskrit) The primal source and ultimate goal of all, the one, divine reality, which pervades the manifold world of things and lives and minds. It is infinite and spaceless and timeless and yet the source for space and time. **Atman-project** Ken Wilber's term for the ascent of the human spirit up the spectrum of consciousness to fulfill its destiny and return to its true home, Atman.

Bell's theorem (or inequality) The correlations discovered by physicist John Stewart Bell between simultaneous measurements of two widely separated particles, in which a limit is observed if reality is local (governed by local interactions). Alain Aspect experimentally checked a pair of photons and found the limit violated. Therefore, it is generally agreed that any model of reality must be nonlocal. *See* locality/nonlocality.

big bang theory The theory that the universe was created by a gigantic explosion from a singularity (*see* singularity) about eighteen billion years ago. Its main confirmation is the detection of black-body background radiation left over from the original event. The theory states that there was no pre-existing space or time before the original bang.

black body An object capable of efficiently absorbing all frequencies of electro-magnetic radiation and emitting all frequencies when brought to incandescence. A true black body is a perfect absorber and radiator. Physicist Max Planck discovered the quanta while studying theoretical problems connected with black-body radiation.

black hole A material body that has collapsed to a highly compressed state in which the gravitational field is so intense that not even light can escape from its surface. It is characterized by only three properties: mass, charge, and angular momentum (although Stephen Hawking has shown that quantum effects do permit some radiation). **white hole** A hypothetical time-reversed black hole. While a black hole absorbs all matter, a white hole is a source for all matter. Mathematically it is possible that a black hole is connected to a white hole through a tunnel that may end up in another part of the universe or in a second universe.

causality Interrelationship between cause and effect. If an effect is a necessary result of a given cause, then causality is operating. However, causality must account for *all* factors governing the change. Since the view presented here is that all systems (including the universe) are abstractions from a greater whole, then all factors are not known, and causality must be conditional.

collective unconscious Jungian term for universal, shared aspect of the unconscious mind, which contains patterns of instinctual behavior (archetypes). The collective unconscious is manifest in the personal unconscious in the forms of images and instincts.

consciousness In this text, a particular quality of being, an infinite spectrum of layers which are not separate but mutually interpenetrated. Bohm refers to these layers as implicate orders, Wilber as levels and Seth as frameworks. All would agree that each individual layer is an aspect of an underlying whole and all layers are accessible with the proper focus. We are all on a journey to widen our focus.

consciousness unit (CU) Seth's term for the smallest units of consciousness — "unit" being a term of convenience, since according to Seth energy is not divided. CUs have their source outside space-time and are not affected by our limits on velocity. They slow down to our velocity of light to create physical structure. They are the source for *all* consciousness and are endowed with unpredictability, which allows for infinite patterns and fulfillments.

coordinate points In Seth's concept, points where other realities coincide with our own and which act as channels through which energy flows; invisible paths from one reality to another. In a manner similar to that in which electrical transformers change the volt-ampere relationship, coordinate points transform thoughts and emotions into matter. There are three kinds of coordinate points: absolute, main, and subordinate.

determinism The philosophical doctrine that assumes that if the state of a system is known at a given time, and if the physical laws governing its behavior are known, all future states can be determined, at least in principle. Most physical theories prior to quantum theory have been deterministic in this sense.

deep structure The defining form of each level of consciousness in Ken Wilber's hierarchy. Each level is divided into a deep structure and surface structure. The deep structure contains all the potentials and limitations of a given level. The surface structure is a particular form of the deep structure; within the potentials and limitations of the deep structure, the surface structure can assume any pattern.

dimensional levels *See* spectrum of consciousness.

EE units Units of energy, which, according to Seth, emanate from all gestalts of consciousness. EE units are the electromagnetic expression of subjective experience. They are emitted by consciousness and propelled into physical actualization if the intensity of the feeling, thought, or emotion from which they arise is of sufficient strength. If not, they can be seen as latent matter. EE units exist at many velocities, but they are perceived by us only at our velocity of light.

ego (individual consciousness) Seth calls our normal waking consciousness, which operates in the three-dimensional world, the outer ego. The inner ego is a higher level of consciousness, which creates the physical world from the inner reality of the collective unconscious. However, there is an even higher level of consciousness which Seth calls your entity or soul. The inner and outer ego are projections from this higher level. There are no real divisions between the levels, but the idea is useful to convey this difficult concept.

entropy In statistical mechanics, a measure of disorder in a closed system. Disorder refers to the evenness of the distribution of energy: an increase in evenness is an increase in disorder. According to the second law of thermodynamics, entropy never decreases in a *total* system. If the universe is an isolated system, it perpetually increases in disorder.

epiphenomenon A phenomenon that arises from the organization of a particular system but is not present in its constituent parts.

events Points in four-dimensional space-time, which do not exhibit extension or duration. In the Newtonian view, the universe consists of things, whereas in the Einsteinian view, the universe consists of events, e.g., the interaction of subatomic particles.

evolution Development or growth that produces higher and higher wholes; a movement from the lower to the higher. In the Perennial Philosophy, evolution is toward the highest state of being, or Atman. *See* involution, microgeny.

explicate order Bohm's term for the domain referred to by Cartesian coordinates (locality in space-time). It displays the separateness and independence of fundamental constituents and is manifest or visible (directly or with instruments). It is secondary to the implicate order (*see below*), which unfolds to create the explicate order and enfolds to give guidance to itself.

field An area of space characterized by a physical property that is normally invisible and intangible but under certain circumstances can interact with matter. Classically, the action between two material objects separated in space is described in the language of fields. In quantum field theory, the interaction is viewed as an exchange of so-called messenger particles, which conveys a particular force. Examples are the photon (electromagnetic) and graviton (gravitational).

Fourier analysis A mathematical technique for analyzing a periodic function (e.g., a musical note) into its fundamental and harmonic components.

frame of reference A set of coordinate axes by which the position or location of an object may be specified.

frameworks Seth's term for interpenetrating levels of action or existence; analogous to the spectrum of consciousness. Our outer reality is Framework 1, and our inner reality and the creative source from which we form events is Framework 2. We draw from Framework 2 by focusing our attention properly. All That Is contains an infinite number of frameworks.

ground unconscious Ken Wilber's term for all the deep structures (*see above*) of all levels of consciousness existing in potential form, which eventually emerge as a particular surface structure within the constraints of the deep structure of that level.

hidden variables Variables that would allow physicists to define the quantum state more precisely than present quantum theory allows. If such variables exist, they probably will be discovered on the subquantum level where present quantum theory may be incomplete. Hidden variables would select among the given probabilities of the wave function.

Hilbert space The abstract space used by physicists to describe quantum mechanical states. A single point in Hilbert space represents the entire quantum system. Unlike the classical concept of phase space, Hilbert space is a *complex* vector space. The wave function is represented by a vector (state) composed of a superposition of states, each corresponding to a possible result of a measurement.

hologram A photographic record created by laser light directly reflected from an object and laser light going directly to the plate to form a complex interference pattern. When the plate is illuminated with coherent light, the wavefront produced creates a three-dimensional image of the original object. In this way, the whole image can be recreated from each portion of the record.

holomovement In David Bohm's terminology, the total ground of that which is manifest. The manifest is embedded in the holomovement, which exhibits a basic movement of unfolding and enfolding.

hyperspace Space that is defined by more than three dimensions. As in Euclidean space, all axes are normal to each other.

implicate order The basic order, according to Bohm, from which our three-dimensional world springs. It is multidimensional, and its connections are independent of space and time. The implicate order is identified with the wave function in quantum theory. *See* explicate order.

involution In Ken Wilber's terminology, the movement of Atman down through the spectrum of consciousness to create the manifest world; the process by which the higher levels of being are involved with lower levels. *See* microgeny.

karma The collective force of a person's actions. According to Indian philosophy, karma persists through reincarnations and determines the content of each lifetime.

light In this book, light refers to the entire electromagnetic spectrum. It can be interpreted as a wave, as in classical theory, or as a stream of quanta or photons, as in quantum theory. The frequency of the wave is related to the energy of the photon by Planck's constant. The velocity of light in free space is constant regardless of the motion of the observer.

Bohm sees the implicate order as an ocean of energy, or light. Matter is a condensation of this energy; it arises when light rays reflect back and forth to form a pattern. Without reflection, there is simply pure light. Matter carries with it time and space, but pure light does not.

Seth states that the electromagnetic spectrum that we observe is a small portion of an infinite spectrum. The attribute that separates sections of the spectrum is the velocity of light. For light to become manifest as physical form, its velocity must be increased or decreased to our range (300,000 km/sec).

linearity/nonlinearity The relationship between input and output in a system. In linear systems the input is always directly proportional to the output. Such systems can be separated into parts so that the whole is exactly equal to the sum of the parts. In nonlinear systems, a given input can cause unexpected outputs, and the whole can be greater than its parts.

locality/nonlocality The condition that defines the causal relationship between events. In a local reality, information cannot travel faster than light. In a nonlocal reality, objects can influence each other instantaneously. Bell's theorem (*see above*) indicates that if the world is made up of separate objects, they must have nonlocal connections. Bohm's concept of quantum potential supports such connections.

matter According to Bohm, Wilber, and Seth, matter is condensed consciousness. The unfolding and enfolding process creates successive localized manifestations that appear to our senses and instruments as physical form.

mechanism The philosophy that envisions the universe as composed of basic entities that follow an absolute set of quantitative laws. The mechanistic view is that if these laws were known and the intellect were large enough, all future events could be calculated and predicted with complete precision. This idea emerged from an extrapolation of Newtonian mechanics to all possible domains of knowledge.

metaphor A word, phrase, or description applied to something that it does not literally designate, thereby suggesting comparison or analogy. For example, the Newtonian universe is often referred to as a machine.

microgeny In Ken Wilber's interpretation of the Perennial Philosophy, the moment-to-moment involution of the spectrum of consciousness. *See* involution.

moment point Seth's term for the present moment, the point of interaction between all existences and realities, through which all probabilities flow. According to Seth, the past, present, and future exist together and can be experienced in a moment. This is analogous to Minkowski's space-time in which all points of space and all points of time exist en bloc. The moving present we perceive as time is, in Seth's concept, a projection from a higher-dimensional timeless order.

mystic One who is capable of transcending the physical realm of space and time through a state of consciousness that extends beyond the restrictions of the intellect to unity with the absolute.

paradigm In science, a conceptual framework, endorsed by the large majority of the scientific community, that presents problems to solve and defines boundaries within which solutions are sought. The term is often applied to the societal realm, in which it refers to the concepts and values of a community that shape its perception of reality.

Perennial Philosophy The term popularized by Aldous Huxley to refer to the consensus of mystics from many ages and cultures that a transcendental unity is the reality of all things.

plenum Space that is completely filled with matter/energy; the opposite of vacuum.

quantum The smallest discrete unit of energy released or absorbed in a process. The quantum for electromagnetic energy is the photon.

quantum field theory The theory that results from the application of quantum mechanics to the behavior of a field. Classically, all fields in physics were treated as continuous, extending to all points of space. However, when quantum mechanics was applied to the field, it became quantized and exhibited definite energy states. This allowed for a particle interpretation and made the wave-particle duality somewhat more understandable.

quantum potential According to David Bohm, a potential that acts on a sub-atomic particle in addition to normal classical potentials. When two particles are separated in space, unconnected by any classical potentials, they are still correlated through the quantum potential, which acts instantaneously. Instantaneous action violates the spirit of relativity, at the least. It is through the quantum potential that Bohm arrives at his concept of wholeness wherein the two particles are not seen as separate entities but as projection from a higher-dimensional domain.

quantum theory A mathematical theory that describes the behavior of all systems (in principle) but is especially useful in the atomic and subatomic realms. According to quantum theory, in*dividual* events are inherently unpredictable, nonseparable, and in the view of some physicists, observer-dependent.

reductionism The principle that seeks to explain complex phenomena in terms of simpler ones. In the reductionist view, psychology can be explained by biology, which is reduced to chemistry and ultimately to particle physics. Reductionism denies the possibility of a collective property that supercedes and cannot be explained by the component parts.

relativity theory, general: The theory that explains the difference between accelerated and nonaccelerated systems and its relationship to gravity. As a consequence, space-time is seen not as rigid but as elastic, with its curvature determined by matter and energy, and gravity becomes a product of geometry rather than a force. These aspects of relativity theory resulted in the concepts of black holes, finite but unbounded universes, and even rips in space-time. **special:** The theory that postulates that the velocity of light is constant for all observers or sources. Also, physical laws are the same in all inertial frames of reference. From these assumptions, Einstein deduced the equivalence of mass and energy and the elasticity of space and time.

Schrödinger's equation An equation that describes the time evolution of a quantum state as represented by a quantum wave function (*see below*). It is a complex differential equation which is completely deterministic but when a measurement is made, the quantum state is abruptly changed to one of a set of new possible states for which the probability of occurrence can be computed from the wave function.

singularity A boundary to the three-dimensional universe where the laws of physics may not be operable; e.g., the center of a black hole, which is infinitely dense and thereby has literally ripped a hole in space-time.

space-time A four-dimensional mathematical construction, based on the work of Minkowski and Einstein, which describes our world more accurately than the three-dimensional concept. This construction arose naturally from relativity's length contraction and time dilations at high velocities. What we call space *and* time can be understood as projections from the space-time world, and they vary according to the velocity of the object.

spectrum of consciousness The hierarchy of increasing dimensions of consciousness, according to the Perennial Philosophy. The greater the number of dimensions, the higher the level. The lowest level is the most dense and fragmentary (matter and energy), the highest level the most subtle and unitary (All That Is).

state vector A vector in abstract Hilbert space that characterizes a quantum mechanical state. It is similar to a wave function but is expressed in quite different mathematical terms.

superimplicate order A super-information field that makes the implicate order designated by the quantum wave function nonlinear and thereby organizes into relatively stable forms (explicate order). It is also a wave function or a super-wave-function.

uncertainty principle The statement by Werner Heisenberg that it is impossible to measure a pair of noncompatible observables (e.g., position and momentum) both precisely and simultaneously because all energy comes in quanta and all matter has both wave and particle aspects. An accurate measurement of position requires a short wavelength (high energy), and an accurate momentum measurement requires long wavelength (low energy).

virtual particle A quantum particle that appears and disappears spontaneously and whose existence is of an extremely short duration. The Heisenberg uncertainty principle allows for this phenomenon, and as a result these short-lived particles are considered to be different from the more familiar observable particles.

wave frequency The number of complete cycles per unit time of a vibrating system.

wave function A mathematical description of the state of a quantum system; a solution of Schrödinger's wave equation.

wavelength The distance between two successive crests of one wave.

wave packet The configuration that results when a wave function is confined to a small region of space. This implies that the particle being described is confined to a fairly localized position.

wave-particle duality The twofold nature of the quantum objects, which can display wavelike and particlelike properties.

white hole *See* black hole.

wholeness Bohm's concept of an undivided, flowing movement, unbroken and all-encompassing but not complete and static. Division or analysis into an arrangement of objects and events can be appropriate for a limited description or display but is not the primary reality or ground.

zero-point energy The irreducible quantity of energy, even at the minimum energy possible, of the discrete states of the quantum field; a result of the uncertainty principle.

Index

D

E

F

G

H

I J K L

Seth Network International (SNI) is a network of
Seth/Jane Roberts readers from over 30 countries
who meet to explore ideas from the Seth material.

For further information, please contact:
Seth Network International
P.O. Box 1620
Eugene, OR 97440 USA
Phone (541) 683-0803 Fax (541) 683-1084
http://www.efn.org/~sethweb